Springer Theses

Recognizing Outstanding Ph.D. Research

Aims and Scope

The series "Springer Theses" brings together a selection of the very best Ph.D. theses from around the world and across the physical sciences. Nominated and endorsed by two recognized specialists, each published volume has been selected for its scientific excellence and the high impact of its contents for the pertinent field of research. For greater accessibility to non-specialists, the published versions include an extended introduction, as well as a foreword by the student's supervisor explaining the special relevance of the work for the field. As a whole, the series will provide a valuable resource both for newcomers to the research fields described, and for other scientists seeking detailed background information on special questions. Finally, it provides an accredited documentation of the valuable contributions made by today's younger generation of scientists.

Theses are accepted into the series by invited nomination only and must fulfill all of the following criteria

- They must be written in good English.
- The topic should fall within the confines of Chemistry, Physics, Earth Sciences, Engineering and related interdisciplinary fields such as Materials, Nanoscience, Chemical Engineering, Complex Systems and Biophysics.
- The work reported in the thesis must represent a significant scientific advance.
- If the thesis includes previously published material, permission to reproduce this must be gained from the respective copyright holder.
- They must have been examined and passed during the 12 months prior to nomination.
- Each thesis should include a foreword by the supervisor outlining the significance of its content.
- The theses should have a clearly defined structure including an introduction accessible to scientists not expert in that particular field.

More information about this series at http://www.springer.com/series/8790

Chris De Gruyter

Travel Plans for New Residential Developments: Insights from Theory and Practice

Doctoral Thesis accepted by
Monash University, Clayton, Australia

Springer

Author
Dr. Chris De Gruyter
Department of Civil Engineering
Institute of Transport Studies
Monash University
Clayton, Victoria
Australia

Supervisor
Prof. Geoffrey Rose
Department of Civil Engineering
Institute of Transport Studies
Monash University
Clayton, Victoria
Australia

ISSN 2190-5053 ISSN 2190-5061 (electronic)
Springer Theses
ISBN 978-981-10-2091-9 ISBN 978-981-10-2092-6 (eBook)
DOI 10.1007/978-981-10-2092-6

Library of Congress Control Number: 2016946328

Printed on acid-free paper

This Springer imprint is published by Springer Nature
The registered company is Springer Science+Business Media Singapore Pte Ltd.

Supervisor's Foreword

Transport challenges are evident in cities across the world. The combination of the growing world population, compounded by the clear preference for the majority of people to live in cities, is magnifying those challenges. Cities faced with limited physical space for new transport infrastructure are struggling with rising traffic congestion, declining air quality, and a deteriorating quality of life for their residents.

While a 'predict and provide' approach to transport planning was the predominant paradigm for many years, there is now a clear recognition of the role that demand-side strategies can play. One such strategy of increasing interest is the travel plan, a mechanism used for delivering a set of transport measures to manage car use and promote a greater uptake of public transport, walking and cycling.

This book explores the use of travel plans in the context of new residential developments, a topic worthy of much attention but one that was largely unchartered. The book provides coverage of the scale of travel planning practice for new developments, industry perspectives on their development and implementation, and an assessment of their quality and effectiveness. The theoretical foundation for the research draws on implementation theory and planning enforcement theory. Those theories are applied in a practical manner to identify opportunities to enhance the impacts of travel plans for new residential developments.

This book finds that while travel plans can be associated with lower car use at new residential developments, a number of opportunities can be realised to enhance their effectiveness. A valuable contribution is provided through the development of an integrated theory of implementation and enforcement as a framework to guide future travel planning practice.

Another key contribution provided by this book is its exploration of 'self-selection' in the context of travel plans for new residential developments. The phenomenon can occur when residents choose to 'self-select' into a new development with a travel plan because it is consistent with their attitudes and preferences towards more sustainable travel. A better understanding of self-selection issues, as this book provides, is critical in evaluating the effectiveness of residential travel plans.

This book employs a detailed and rigorous methodology for meeting its research objectives and presents a clearly considered set of results and conclusions. It should provide a valuable resource to transport researchers and those professionals involved in the planning of new residential developments.

Victoria, Australia
July 2016

Prof. Geoffrey Rose

Abstract

Continued demand for new housing development is expected to add further pressure to existing transport networks and services in many urbanised areas. Given these challenges and a limited ability to add more capacity to the transport network, it is appropriate to consider the role of demand-side strategies, such as travel plans. Travel plans aim to manage car use among building occupants by providing a package of site-specific initiatives and facilities that support access by more sustainable forms of transport. They can be required through the land use planning and approval process for new and expanded buildings, such as offices, schools and residential developments. However, there is a limited understanding of the effectiveness of travel plans when applied to new residential developments. Furthermore, the implementation of travel plans at new residential developments has not been sufficiently explored.

This thesis aims to assess the effectiveness of travel plans for new residential developments and identify opportunities to enhance their effectiveness. A mixed methods approach comprising five key research components is adopted to achieve this aim, including the application and integration of both implementation theory and planning enforcement theory.

The first component involves a survey of councils to examine the scale of travel planning practice for new urban developments in Victoria, Australia. Results show that half of the councils had previously required a travel plan for a new development, primarily to offset the impact of less car parking space being provided. Around 100 travel plans were found to be required during 2010–2012 alone, yet 80 % of councils had not monitored any of those travel plans.

The second component develops an appreciation for the perspectives of industry actors involved in travel planning for new residential developments through a set of interviews. This shows general support for travel plans at new residential developments, but limited confidence in the ability to implement them successfully. Implementation challenges were found to centre on a lack of enforcement, uncertainty regarding implementation responsibilities, and a general of lack of ownership of travel plans when applied to residential settings.

The third component provides an assessment of the quality of travel plans prepared for new residential developments against a best practice framework. This shows considerable scope to improve travel plan quality, particularly in estimating expected travel patterns of future building occupants, specifying how the travel plan will be managed and implemented, and outlining clearer processes for monitoring and review.

The fourth component provides an assessment of the effectiveness of travel plans at new residential developments. A set of multi-modal trip counts reveal that car use at new residential developments with travel plans was 14 percentage points lower than matched control sites. In addition, it provides some preliminary evidence of residents 'self-selecting' into developments with travel plans, with this accounting for a relatively small yet non-trivial proportion of observed differences in travel behaviour.

The fifth and final component views the research findings through the lens of both implementation theory and planning enforcement theory to identify opportunities to enhance the effectiveness of travel plans for new residential developments. Short-term enhancements include greater ownership and engagement of 'implementers', improvements to travel plan quality, provision of guidance material and training, and a more pro-active and facilitative style of enforcement. Long-term enhancements include sound planning requirements, a stronger industry focus for residential travel planning and ensuring an adequate number of technically competent staff are available for enforcement. An integrated theory of implementation and enforcement, with consideration to both top-down and bottom-up styles of implementation, and both facilitative and systematic styles of enforcement, is developed to guide future travel planning practice.

This thesis provides a number of original contributions to knowledge in the field of travel planning for new residential developments. Overall, it is concluded that while travel plans can be effective in reducing car use at new residential developments, a number of opportunities can be realised to enhance their effectiveness. Acting on these opportunities will require sufficient resources and commitment. However, this will ultimately improve the way in which travel plans are developed, implemented and monitored at new residential development into the future, thereby supporting a greater uptake of more sustainable forms of transport.

Parts of this thesis have been published in the following journal articles:

De Gruyter, C, Rose, G & Currie, G (In Press) 'Travel Plans for New Residential Developments: Measuring Self-Selection Effects to Better Understand Travel Behaviour Impacts', *Transportation Research Record: Journal of the Transportation Research Board*, no. 2564 (Accepted for publication 26th February 2016).

De Gruyter, C, Rose, G & Currie, G (2015) 'Understanding Travel Plan Effectiveness for New Residential Developments', *Transportation Research Record: Journal of the Transportation Research Board*, no. 2537, pp. 126–136.

De Gruyter, C, Rose, G & Currie G (2015) 'Enhancing the impact of travel plans for new residential developments: Insights from implementation theory', *Transport Policy*, vol. 40, pp. 24–35.

De Gruyter, C, Rose, G & Currie G (2014) 'Methodology for Evaluating Quality of Travel Plans for New Developments', *Transportation Research Record: Journal of the Transportation Research Board*, no. 2417, pp. 46–57.

De Gruyter, C, Rose, G & Currie G (2014) 'Securing travel plans through the planning approvals process: A case study of practice from Victoria, Australia', *Cities*, vol. 41 part A, pp. 114–122.

Acknowledgements

This thesis would not have been possible without the support from a number of people. Sincere thanks go to both my supervisors, Prof. Geoff Rose and Prof. Graham Currie, for their excellent guidance and advice, but also for developing my skills as an independent researcher. Words cannot express how fortunate I feel to have had the opportunity to be supervised by such talented individuals. Prof. Bill Young is also acknowledged for his role as an informal mentor and for all of the interesting conversations we had along the way. The Ph.D. students in the transport group are also thanked for making this journey an extremely enjoyable one, always reminding me to never take things too seriously. In particular, I would like to express my thanks to Brendan Pender (I will dearly miss our coffee runs!) and Kelvin Goh (one of the most humble people I have ever met). I am also grateful to Jenny Manson, our Research and Postgraduate Manager; how she can make each Ph.D. student feel as though they are the only one in the department (when there are actually more than 100 of us!) never ceases to amaze me. Many thanks also go to David Meiklejohn who first provided me with the opportunity to get involved with travel plans back in 2004, and for his continued advice and friendship.

Each of the council representatives and interview participants are thanked for their assistance with participating in this research. In particular, I would like to thank James Noy (Public Transport Victoria), Richard Armitage (RATC, United Kingdom) and Chrisanthy Semertzidis (City of Stonnington). Also acknowledged are each of the property and building managers who enthusiastically supported the research. Brenda O'Keefe from the Institute of Transport Studies is also acknowledged for her brilliant assistance with the production of survey material, as are the 16 civil engineering students for their tireless efforts in supporting the data collection.

Most importantly, I would like to thank my wife Elaine who now knows a lot more about travel plans than she probably ever wanted to, who could not feel her toes on many cold mornings while helping me count cars, who patiently listened to every presentation I practised, and who read over every word in this thesis. I would also like to thank my mother for her continual support and for providing such helpful comments on each draft chapter of this thesis. Other family members—Melissa, Andrew, Kylie, Paris and Damien—are also thanked for offering a fresh set of perspectives and showing such a keen interest in my research topic.

Contents

Abbreviations

ACF	Advocacy Coalition Framework
ATE	Average Treatment Effect
ATTrBuTE	Assessment Tool for Travel Plan Reviewing, Building, Testing and Evaluation
AVR	Average Vehicle Ridership
BedZED	Beddington Zero Energy Development
BUG	Bicycle User Group
CBD	Central Business District
CTR	Commute Trip Reduction
EPA	Environment Protection Authority
ESD	Environmentally Sustainable Design
ICC	Intraclass Correlation Coefficient
ITE	Institute of Transportation Engineers
MUHREC	Monash University Human Research Ethics Committee
NHS	National Health Service
OD	Observed Difference
OECD	Organisation for Economic Cooperation and Development
PSM	Propensity Score Matching
QR	Quick Response
RMS	Roads and Maritime Services
RTA	Roads and Traffic Authority
SMART	Specific, Measurable, Achievable, Relevant, Time-based
SSE	Self-Selection Effect
TAFE	Technical and Further Education
TDM	Travel Demand Management
TRO	Trip Reduction Ordinance
UK	United Kingdom
URL	Uniform Resource Locator
US	United States
VISTA	Victorian Integrated Survey of Travel and Activity

List of Figures

List of Tables

Chapter 1
Introduction

1.1 Overview

A travel plan can be defined as a strategy containing a package of tailored initiatives and facilities delivered at a site to manage car use and encourage the use of more sustainable forms of transport, such as walking, cycling and public transport (Enoch 2012).

This thesis explores the use of travel plans for new residential developments. It aims to provide an understanding of their effectiveness, and draws upon a theoretical model to identify opportunities to enhance their effectiveness and guide future practice.

This introduction provides the background and motivation for this research, followed by a description of its aim and objectives. The scope, theoretical context, and contributions of the research are also discussed. The chapter concludes with an outline of the thesis structure.

1.2 Background and Motivation

Major cities across the world are experiencing various transport challenges associated with a growing population, dependency on the motor vehicle, and concerns about the environment (Stopher and Stanley 2014). In Australia, congestion costs arising from increased travel times, higher vehicle operating costs and poorer air quality are expected to rise from \$9 billion in 2005 to \$20 billion per annum by 2020 (Bureau of Transport and Regional Economics 2007). Population growth in Australia has continued to outpace that of all 34 member countries of the Organisation for Economic Cooperation and Development (OECD), at 1.5 % per annum between 2002 and 2012 (OECD 2014). By 2075, Australia's population is predicted to double to 46 million people (Australian Bureau of Statistics 2013).

C. De Gruyter, *Travel Plans for New Residential Developments: Insights from Theory and Practice*, Springer Theses,
DOI 10.1007/978-981-10-2092-6_1

Approaches to dealing with increased pressures on the transport system, particularly population growth, have traditionally focused on providing additional infrastructure to cater for predicted levels of future travel demand (Stopher and Stanley 2014). However, both physical and financial constraints pose challenges in continuing with the 'predict and provide' approach. It is therefore appropriate to consider the role that complementary 'demand side' strategies can play, such as Travel Demand Management (TDM).

TDM is defined by the Institution of Engineers Australia (1996) as 'intervention (excluding provision of major infrastructure) to modify travel decisions so that more desirable transport, social, economic, and/or environmental objectives can be achieved and the adverse impacts of travel can be reduced'. Examples of TDM strategies range from travel awareness programs and staggered working hours to fuel taxes and congestion pricing (Wayte 1991).

In recent years, the use of the 'travel plan' as a TDM strategy has been adopted to assist in managing car use by delivering a package of site-specific initiatives and facilities at key trip generators, such as workplaces and schools (Cairns et al. 2004). The initiatives and facilities contained in a travel plan can be wide-ranging, although the more effective ones tend to include both 'carrots', such as financial incentives to use public transport, and 'sticks', such as car parking limitations to discourage car use (Cairns et al. 2010). More recently, travel plans have been required through the land use planning and approvals process for new and expanded buildings, such as offices, schools and residential developments (Rye et al. 2011a). The use of travel plans specifically for new residential developments is the focus of this thesis.

Travel plans have been required for new residential developments in the United States, United Kingdom and other parts of Europe (Jollon 2013; Rye et al. 2011b), yet little research has been undertaken in this field to date. In general, most research undertaken into travel plans has focused on pre-existing sites such as workplaces and schools (Cairns et al. 2004).

Two key research gaps emerge from the literature review presented in Chap. 2 of this thesis. Firstly, there is an insufficient understanding of the effectiveness of travel plans in reducing car use at new residential developments. Secondly, the implementation of travel plans at new residential developments has not been sufficiently explored; this is particularly relevant given their characteristics when compared to traditional workplace and school travel plans. This thesis aims to address these two key research gaps.

1.3 Research Aim and Objectives

In response to the research gaps identified, the aim of this research is:

To assess the effectiveness of travel plans for new residential developments and to identify opportunities to enhance their effectiveness

In the context of this thesis, 'effectiveness' refers to the ability of the travel plan to reduce car use among residents living at a given development, relative to not having a travel plan in place.

A number of specific research objectives have been identified as key steps required to meet the research aim. These research objectives are framed in response to the wider research gaps identified from the literature review (presented in Chap. 2). In the context of travel plans for new residential developments, the research objectives are:

1. To examine the scale of practice in Victoria, Australia
2. To gain an appreciation for the perspectives of industry actors involved in their application
3. To evaluate their quality and effectiveness
4. To identify and assess opportunities for enhancing their implementation.

1.4 Scope and Theoretical Context

This research is focused specifically on travel plans for new residential developments required through the land use planning and approvals process. The research has been conducted in the Australian state of Victoria. Victoria is located in the south east corner of the Australian mainland and is home to over five million people. The capital city of Victoria is Melbourne, with a population of around four million people (Department of Transport, Planning and Local Infrastructure 2014).

Melbourne has been consistently ranked as the world's most liveable city out of 140 cities surveyed since 2011 (The Economist Intelligence Unit 2014). Infrastructure, which accounts for 20 % of the ranking, takes into account the quality of transport networks and availability of good quality housing (The Economist Intelligence Unit 2014). Melbourne has one of the largest streetcar (tram) networks in the world (Currie and Burke 2013), yet car travel accounts for 65 % of the journey to work by Melbourne residents (Australian Bureau of Statistics 2011).

The findings of this research have wider geographical implications than Melbourne (and Victoria) given that travel plans are used in other states and countries. However, the local context should always be considered.

A mixed methods research approach was adopted which comprised the following key elements:

- **Industry survey**: to gauge the scale of travel planning practice for new urban developments among Victorian councils, providing context for subsequent components of the research
- **Interviews**: to provide insight into the perspectives of industry actors involved in travel planning for new residential developments, particularly issues related to implementation

- **Document reviews**: to provide an assessment of the quality of travel plans prepared for new residential developments against a best practice framework
- **Case studies**: to provide an assessment of the effectiveness of travel plans in reducing car use at new residential developments.

In order to identify opportunities to enhance the implementation and subsequent effectiveness of travel plans for new residential developments, the research draws upon both implementation theory and planning enforcement theory. Implementation theory provides valuable guidance for the effective implementation of programs and policies (Sabatier and Mazmanian 1980), while planning enforcement theory suggests suitable approaches for achieving planning compliance (Burby et al. 1998). These theories therefore provide direct relevance to the implementation and enforcement of travel plans for new residential developments.

This thesis uses the term ‘travel plan’. However, travel plans are also referred to as green travel plans, mobility management plans, trip reduction plans and TDM plans (Enoch and Rye 2006). In addition, the term ‘council’ is used in this thesis to refer to a local government elected authority. Slightly different terms may be used in other countries such as municipal government, local authority or county government. Finally, the term ‘new development’ is used to refer to any new or expanded building (e.g. office, school, residential).

1.5 Contribution of This Study

In line with the scope, this thesis makes **five original contributions to knowledge**. In the context of travel plans for new residential developments, these include:

1. An understanding of the scale of practice in Victoria
2. An appreciation for the perspectives of actors involved
3. An understanding of their quality
4. An understanding of their effectiveness
5. An understanding of how implementation can be enhanced to improve outcomes.

This thesis also provides an important **theoretical contribution** by integrating implementation theory and planning enforcement theory. This helps to facilitate an improved understanding of implementation and enforcement in the context of travel plans for new residential developments.

1.6 Structure of This Thesis

Figure 1.1 presents the structure of this thesis and identifies where original contributions to knowledge flow from the research. The thesis has ten chapters, including this introduction.

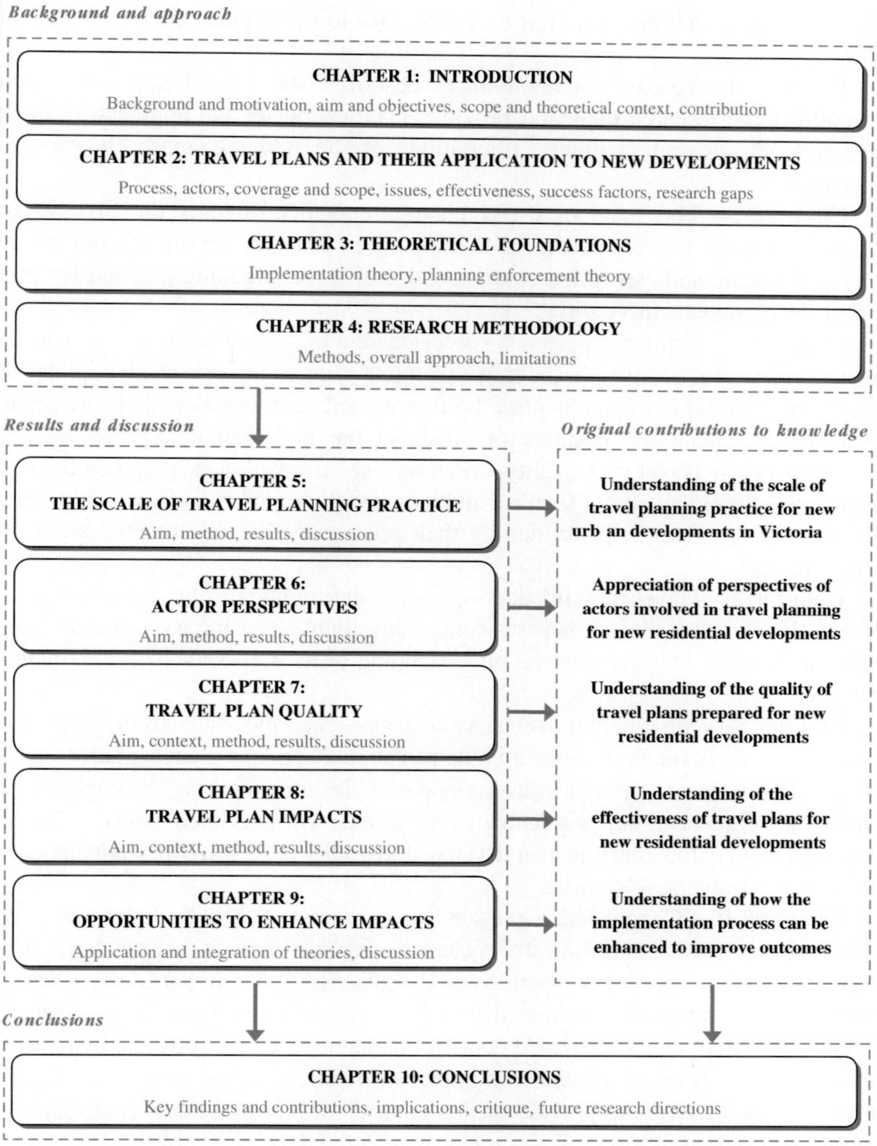

Fig. 1.1 Thesis structure

Chapter 2—Travel plans and their application to new developments—provides a review of the literature on travel plans at both pre-existing sites and new developments, including their application to new residential developments. Research gaps are then identified which are addressed by the thesis in later chapters.

Chapter 3—Theoretical foundations—continues with the literature review by providing an overview of implementation theory and planning enforcement theory, along with a discussion of their relevance to travel plans at new residential developments.

Chapter 4—Research methodology—describes the overall approach taken to address the research gaps and provides justification for the approach through a review of research methods. Limitations of the research approach are also discussed.

Chapter 5—The scale of travel planning practice—details the first set of research results. It presents the findings from an industry survey of councils to gauge the scale and associated characteristics of travel planning practice for new urban developments in Victoria.

Chapter 6—Actor perspectives—describes the findings from a series of interviews with actors involved in the travel planning process for new residential developments. A particular focus is placed on implementation issues and challenges to assist in identifying opportunities for enhancing the implementation process.

Chapter 7—Travel plan quality—reviews a set of travel plans prepared for new residential developments in Victoria and assesses their quality against a best practice framework. This helps to identify their relative merits and potential areas for improvement.

Chapter 8—Travel plan impacts—presents the findings from an evaluation of travel plans implemented at new residential developments using a set of case study sites in Victoria. This provides an understanding of their effectiveness in reducing car use.

Chapter 9—Opportunities to enhance impacts—takes the results from Chaps. 5–8 and views these through the lens of implementation theory and planning enforcement theory. This process assists in assessing opportunities for enhancing the implementation of travel plans at new residential developments. An integrated theory of implementation and enforcement is then developed to guide future travel planning practice for new residential developments.

Chapter 10—Conclusions—presents a summary of key findings and contributions to demonstrate how the research aim and objectives have been met. Implications for theory and practice are also discussed. A critique of the research approach is then provided, followed by a discussion of future research directions.

This introduction has provided context for the research by outlining its motivation, aim, objectives, scope and contributions. The next chapter presents a detailed review of the literature to provide further context for how this thesis contributes to our understanding of travel plans for new residential developments.

References

Australian Bureau of Statistics. (2011). *Census of population and housing data*. Retrieved September 25, 2014, from http://www.abs.gov.au/census.

Australian Bureau of Statistics. (2013). *Population Projections Australia: 2012 (Base) to 2101*. Catalogue number 3222.0, Canberra, Australia.

Burby, R., May, P., & Paterson, R. (1998). Improving compliance with regulations: Choices and outcomes for local government. *Journal of the American Planning Association, 64*(3), 324–334.

Bureau of Transport and Regional Economics. (2007). *Estimating urban traffic and congestion cost trends for Australian cities*. Canberra, Australia: Department of Transport and Regional Services.

Cairns, S., Sloman, L., Newson, C., Anable, J., Kirkbride, A., & Goodwin, P. (2004). *Smarter choices—Changing the way we travel*. UK: Department for Transport.

Cairns, S., Newson, C., & Davis, A. (2010). Understanding successful workplace travel initiatives in the UK. *Transportation Research Part A: Policy and Practice, 44*(7), 473–494.

Currie, G., & Burke, M. (2013). Light Rail in Australia—performance and prospects. Paper presented to 36th Australasian Transport Research Forum (ATRF), Brisbane, Australia.

Department of Transport, Planning and Local Infrastructure. (2014). *Victoria in Future 2014 Data Tables*, Victoria, Australia. Retrieved from http://www.dpcd.vic.gov.au/.

Enoch, M. (2012). *Sustainable transport, mobility management and travel plans*. Surrey, England: Ashgate Publishing Limited.

Enoch, M., & Rye, T. (2006). Travel plans: Using good practice to inform future policy. In B. Jourquin, P. Rietveld, & K. Westin (Eds.), *Towards better performing transport networks* (pp. 157–177). London, UK: Routledge.

Institution of Engineers Australia. (1996). *Policy on travel demand management in urban areas*, Canberra, Australia.

Jollon, M. (2013). *Best practices in integrating travel plans with the development approval process: Examples from the USA*. Paper presented to Planning Institute of Australia (PIA) 2013 National Congress, Canberra, Australia.

OECD. (2014). *OECD Factbook 2014: Economic, environmental and social statistics*. OECD Publishing. Retrieved September 25, 2014, from http://www.oecd-ilibrary.org/economics/oecd-factbook_18147364.

Rye, T., Green, C., Young, E., & Ison, S. (2011a). Using the land-use planning process to secure travel plans: An assessment of progress in England to date. *Journal of Transport Geography, 19*(2), 235–243.

Rye, T., Welsch, J., Plevnik, A., & de Tomassi, R. (2011b). First steps towards cross-national transfer in integrating mobility management and land use planning in the EU and Switzerland. *Transport Policy, 18*, 533–543.

Sabatier, P., & Mazmanian, D. (1980). The implementation of public policy: A framework of analysis. *Policy Studies Journal, 8*(4), 538–560.

Stopher, P., & Stanley, J. (2014). *Introduction to transport policy: A public policy view*. Cheltenham, UK: Edward Elgar Publishing Limited.

The Economist Intelligence Unit. (2014). *A summary of the liveability ranking and overview*. Retrieved December 23, 2014, from http://pages.eiu.com/rs/eiu2/images/Liveability_rankings_2014.pdf.

Wayte, A. (1991). *Road demand management study—a synopsis*. Paper presented to Road Demand Management Seminar, Austroads, Perth, Australia.

Chapter 2
Travel Plans and Their Application to New Developments

2.1 Introduction

The aim of this chapter is to provide a review of the literature on travel plans and their application to new developments. Research gaps identified from the literature review are also discussed which then become the focus of this thesis in subsequent chapters (Fig. 2.1).

To date, research into travel plans has predominantly focused on their application to pre-existing sites, particularly workplaces and schools (Cairns et al. 2004). This has provided valuable guidance for applying travel plans to new developments. For this reason, this chapter draws upon the travel planning literature concerned with both pre-existing sites and new developments.

This literature review also draws upon the experience with developing travel plans for different types of land uses. Examples of land uses where travel plans have been developed include:

- Offices (Cairns et al. 2010)
- Primary and secondary schools (Smith 2010)
- Universities (Curtis and Holling 2004)
- Hospitals (Khandokar et al. 2013)
- Residential sites (Department for Transport 2005)
- Airports (Ison et al. 2014)
- Railway stations (ATOC 2013)
- Retail/shopping centres (Woodruff and Hui 2010)
- Sporting venues and events (Currie and Delbosc 2011)
- Tourist attractions (Guiver and Stanford 2014)
- Mixed use developments (Wiblin 2010).

This literature review does not cover the use of personalised journey planning techniques in the context of household and community based programs. These

C. De Gruyter, *Travel Plans for New Residential Developments: Insights from Theory and Practice*, Springer Theses,
DOI 10.1007/978-981-10-2092-6_2

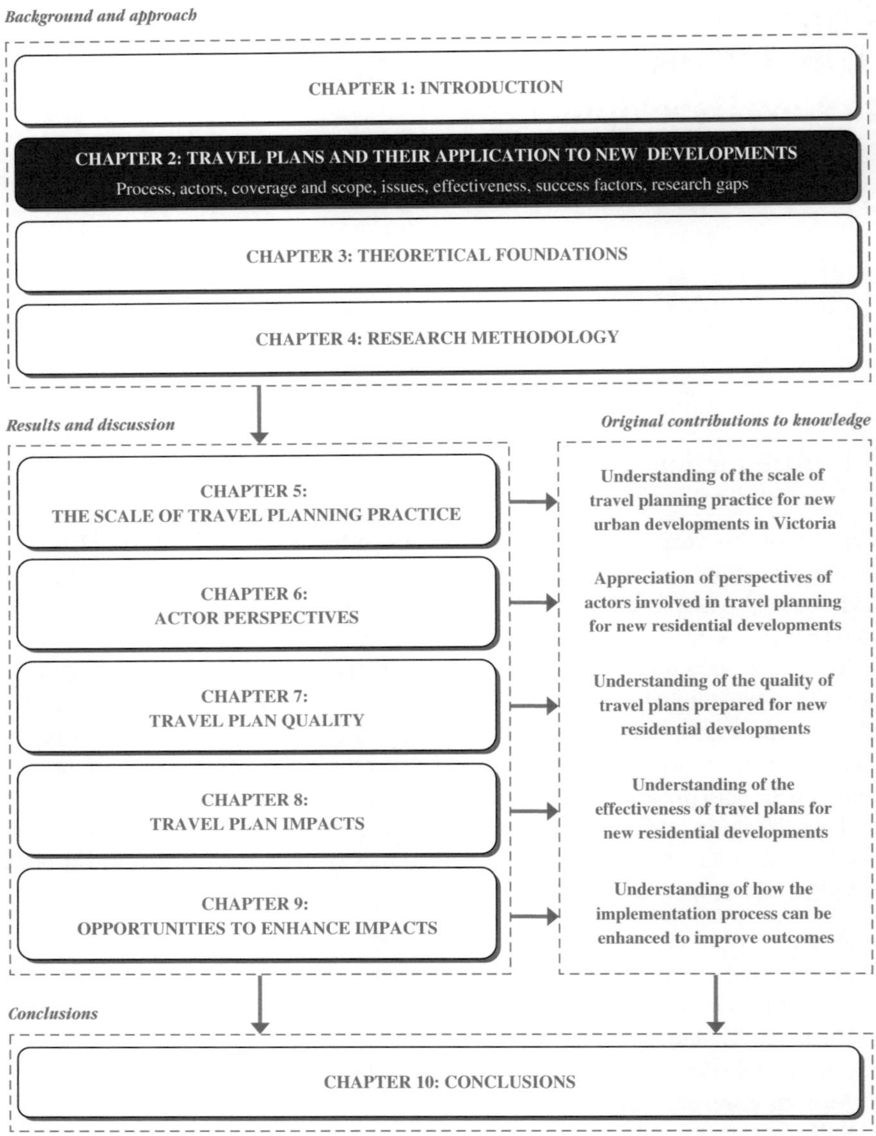

Fig. 2.1 Position of Chap. 2 in the thesis structure

programs have been extensively researched in the past 10–15 years under various names such as individualised travel marketing and travel blending (Brög et al. 2009; Rose and Ampt 2001). They are considered different to travel plans in that they represent a specific voluntary travel behaviour change initiative focused primarily on information, awareness and feedback. This is in contrast to a travel plan which represents a mechanism for delivering a package of travel initiatives over time (Enoch 2012).

In order to gain a comprehensive and holistic understanding of travel plans, the objectives of this literature review are to provide an understanding of:

- The travel planning process
- Actors involved in the travel planning process
- Geographical coverage and scope of travel plans
- Issues with requiring travel plans for new developments
- Methods for evaluating the effectiveness of travel plans and their associated results
- Key success factors for travel plans.

This chapter is structured in line with these objectives. The chapter concludes with the identification of research gaps arising from the literature review and a discussion of the opportunities for addressing these gaps.

2.2 The Travel Planning Process

The travel planning process is commonly described as a series of steps (Howlett and Watson 2010) as follows:

1. **Securing commitment and ownership**: securing support of decision makers, identifying who to involve, confirming objectives and identifying the benefits of developing a travel plan
2. **Understanding the existing situation**: gathering information on current travel behaviour through surveys and/or understanding the existing transport characteristics of the site
3. **Developing the travel plan**: identifying strategies and actions aimed at encouraging travel behaviour change that will form the basis of the travel plan
4. **Implementing the travel plan**: implementing the travel plan and maintaining participation
5. **Monitoring, reviewing and sustaining**: measuring the success of the travel plan and refining aspects where necessary.

These steps are illustrated in Fig. 2.2. The circular nature of the figure illustrates that travel plans are an ongoing process, with the need to revisit previous steps to ensure the travel plan continues to maintain relevance in light of any changing circumstances (Howlett and Watson 2010).

Travel plans are typically implemented by a coordinator, preferably based at the site with support from a working group. Most travel plans also include a set of objectives and targets which are monitored and reviewed on a regular basis (Department for Transport 2009).

However, prior to a travel plan being developed, it may have first been required using a planning condition or formal agreement, in the case of a new development. A formal agreement is considered to have more ‘legal force’ as it can be used to secure payments associated with implementing and monitoring the travel

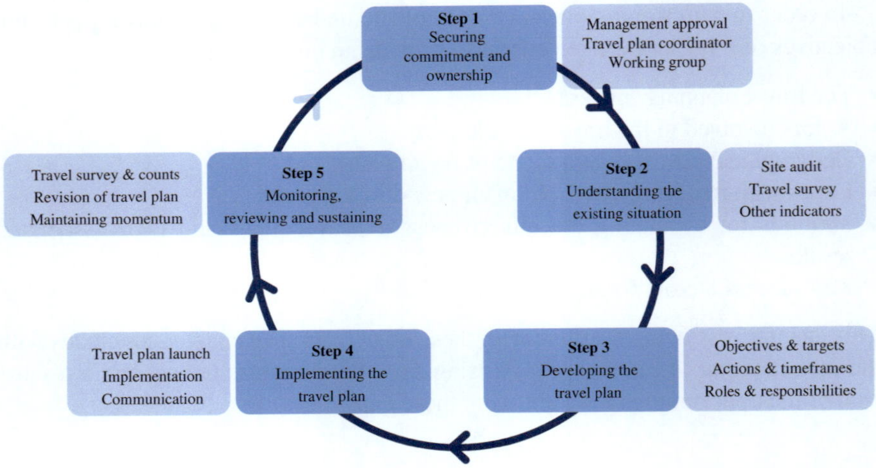

Fig. 2.2 The travel planning process. *Source* Author's adaptation based on Howlett and Watson (2010)

plan (Department for Transport 2002, 2009). It can also be registered over the title of the land and become binding upon future owners (Department of Transport, Planning and Local Infrastructure 2014a). However, Harrison (2003, p. 401) maintains 'there is no need for a belt and braces approach and either conditions or (agreements) concerning travel plans if properly drafted, are perfectly capable of being lawful and enforceable'.

Planning requirements for travel plans are commonly cited as key reasons for their initial preparation (Dill 1998; Roby 2010c; Yeates and Enoch 2012). In particular, Khandokar et al. (2013) found that 98 % of healthcare authorities in the UK cited the requirement as a motivation for developing their travel plan. However, as travel plans develop, other motivations typically sustain them such as car parking supply issues, corporate social responsibility, and the presence of localised congestion and access issues (Roby 2010c).

In developing a travel plan, corresponding to step 3 of Fig. 2.2, guidance available from various countries (ACT Canada & Noxon Associates Limited 2010; Department of Infrastructure 2008; NZ Transport Agency 2011; Transport for London 2011b) generally all recommend that the following elements are included:

- Context
- Existing transport conditions
- Objectives
- Targets and indicators
- Actions/measures
- Management
- Monitoring and review.

The **context** for a travel plan typically includes a summary of site characteristics, such as its physical location and the number of users/occupants (Department of

Infrastructure 2008). An outline of any relevant policies may also be provided at this stage. For example, an employer may have a working from home policy in place which can help in reducing the number of trips to and from the site (Cairns et al. 2004).

Existing transport conditions generally describe the existing transport networks and services in the surrounding area. Where possible, a summary of any relevant local travel survey data helps in understanding existing travel patterns (Transport for London 2011b).

The **objectives** of the travel plan typically set out what the travel plan aims to achieve. **Targets and indicators** are then developed to support these objectives (Department of Infrastructure 2008). While targets and indicators are often focused on outcomes relating to transport mode shares, they can also consider the uptake and awareness of travel plan initiatives and may be framed in both quantitative and qualitative terms (NZ Transport Agency 2011).

The types of **actions/measures** included within travel plans can be wide-ranging. Importantly, these need to be tailored to the site's characteristics and include both demand restraint and positive incentives where possible (Cairns et al. 2010). Appendix A provides a list of over 50 examples of travel plan measures from the literature. These cover the areas of walking, cycling, public transport, car parking, carpooling, car sharing, marketing and promotion, financial incentives, travel plan management, and working practices. However, as noted by Enoch (2012), once contextual constraints are taken into account, the range of measures available to a site may be less than originally envisaged. For example, while rationing car parking may be highly effective in achieving a reduction in car use, the low acceptability of this measure may prevent it from being implemented at many sites. This is supported by a review of over 5,000 workplace travel plans in the United States which showed than only 6 % included parking management measures (Young and Luo 1995).

Management of the travel plan typically stipulates timeframes, roles and responsibilities (including the appointment of a travel plan coordinator) and the available budget (Transport for London 2011b). Handover arrangements may also be specified where the travel plan is prepared a new development and the occupant is not yet known (Department for Transport 2009).

The final element of a travel plan is generally concerned with **monitoring and review**. Here, the frequency, method/s, responsibilities and timing associated with these components is typically described (Department for Transport 2009).

2.3 Actors Involved in the Travel Planning Process

Given that travel plans can be prepared for different types of land uses, a wide range of actors tend to be involved in the travel planning process. Table 2.1 presents a synthesis of the literature relating to key actors and their roles in developing, implementing and monitoring travel plans.

Table 2.1 Actors involved in the travel planning process and their role at key stages

Actor	Development	Implementation	Monitoring
Government agencies			
National government	May provide overarching policy context for travel plans	May provide funding and guidance for implementation	Minimal role; monitoring may reflect national policy goals
State government	Typically provides policy context for travel plans	Typically provides funding and guidance for implementation	May set monitoring standards and guidelines
Local government	Seeks to apply national/state policy at local level	Hands-on role and/or provides guidance for implementation	Lead role or coordination through third party; enforcement
Travel plan organisations			
Workplaces	Typically involved in writing the travel plan	Delivery agent generally responsible for implementation	Lead role or coordination through third party
Schools	Typically involved in writing the travel plan	Delivery agent generally responsible for implementation	Lead role or coordination through third party
Universities	Typically involved in writing the travel plan	Delivery agent generally responsible for implementation	Lead role or coordination through third party
Hospitals	Often involved in writing the travel plan	Delivery agent generally responsible for implementation	Lead role or coordination through third party
Retail centres	Provide input but unlikely to lead process	Some involvement with support from third parties	Provide input but unlikely to lead process
Property managers	May provide input but unlikely to lead process	Generally involved following property occupation	Provide input but unlikely to lead process
Transport users			
Employees	Provide input through survey and/or other forums	Fulfil roles for travel plan coordinator and working group	Provide input through surveys; may assist with monitoring tasks
Students	Provide input through survey and/or other forums	Involved in supporting implementation of initiatives	Provide input through surveys; may assist with monitoring tasks
Teachers	May be involved in writing travel plan	May fulfil roles for travel plan coordinator and working group	May be involved in coordinating monitoring tasks
Parents (for school travel plans)	Provide input through survey and/or other forums	May have representation on working group	Provide input through surveys; may assist with monitoring tasks
Patients (for hospital travel plans)	May be consulted through survey and/or other forums	No role	Provide input through surveys

(continued)

Table 2.1 (continued)

Actor	Development	Implementation	Monitoring
Visitors	May be consulted through survey and/or other forums	No role	Provide input through surveys
Shoppers	May be consulted through survey and/or other forums	No role	Provide input through surveys
Residents	Provide input through survey and/or other forums	May have representation on working group	Provide input through surveys; may assist with monitoring tasks
Others			
Local travel plan groups	Support development through information sharing	May deliver larger initiatives on behalf of several organisations	May coordinate monitoring on behalf of organisations
Property developers	Coordinate and/or write the travel plan	Likely to implement some measures and provide funding	Lead role or provides funding for monitoring
Transport consultants	May prepare travel plans on behalf of organisations	May be engaged as travel plan coordinator	May coordinate monitoring on behalf of organisations
Transport operators	May be consulted about improving/providing services	May deliver new services as part of travel plan	May support monitoring through provision of data/information
Interest groups	May lobby for improvements and comment on travel plan	May provide local guidance and advice where applicable	May request monitoring data to support own interests

Source Author's synthesis of the literature based on Cairns and Newson (2006), Di Pietro and Hughes (2003), Enoch (2012), Holzer (2004), Roby (2010a, c), Woodruff and Hui (2010) and Yeates and Enoch (2012)

Government agencies are likely to be involved in developing policies on travel plans, along with providing funding and guidance to support implementation and monitoring. Local government generally plays a stronger role given the site-specific nature of travel planning. They can also be involved in enforcing travel plans that have been required for new developments.

Travel plan organisations can be wide ranging with workplaces, schools, universities and hospitals having played a strong role to date in all aspects of the travel planning process. Such organisations generally have the most influence in determining the effectiveness of a travel plan in their role as 'delivery agent' (Enoch 2012). However, they are not typically part of the 'traditional' transport policy institutional structure and therefore may only have limited transport knowledge and experience (Enoch 2012).

Transport users, as the recipients or end users of travel plans, are generally involved through participating in surveys and other forums to support the development and monitoring of travel plans. However, they may also be involved in fulfilling the travel plan coordinator role and having representation on a working group, particularly where travel plans are introduced in workplaces and schools.

A number of other actors can also be involved in the travel planning process. In particular, property developers and transport consultants are becoming increasingly involved given requirements in some jurisdictions for travel plans at new developments.

Previous research has sought the perspectives of actors involved in travel planning (Davison et al. 2010; Enoch and Ison 2008; Rye et al. 2011a; Yeates and Enoch 2012), with relevant findings incorporated throughout subsequent sections of this chapter. While Yeates and Enoch (2012) explored the perspectives of developers involved in travel planning for new developments, **no research has specifically explored the perspectives of the different actors involved in travel planning for new residential developments, particularly aspects relating to implementation**.

2.4 Geographical Coverage and Scope

The travel plan concept started in the United States with a focus on carpooling in response to the oil crisis in the 1970s. The concept was later picked up by the United Kingdom and the Netherlands in the 1980s and early 1990s with a focus on reducing car-based commuter trips (Coleman 2000). Today, travel plans have extended their reach into a number of European countries, as well as Australia, New Zealand and Japan (Enoch 2012).

Figure 2.3 illustrates how travel plans have evolved in different countries. The United States and the Netherlands are the only countries with an industry sector in place for travel plan services.

A number of European countries, including the United Kingdom, have a strong level of support for travel plans, while others have less support, such as Australia, or are still in the pilot testing stage. In addition to the countries shown in Fig. 2.3, Singapore has recently commenced their involvement with workplace travel plans, although this is only at a pilot testing stage (Hooi 2012).

The level of detail available in the literature on travel planning activity varies considerably by country. Levels of activity in the United States, United Kingdom and Australia are generally well documented compared to other countries. A synthesis of the literature from these countries, along with others where information was available, is provided in the following sections.

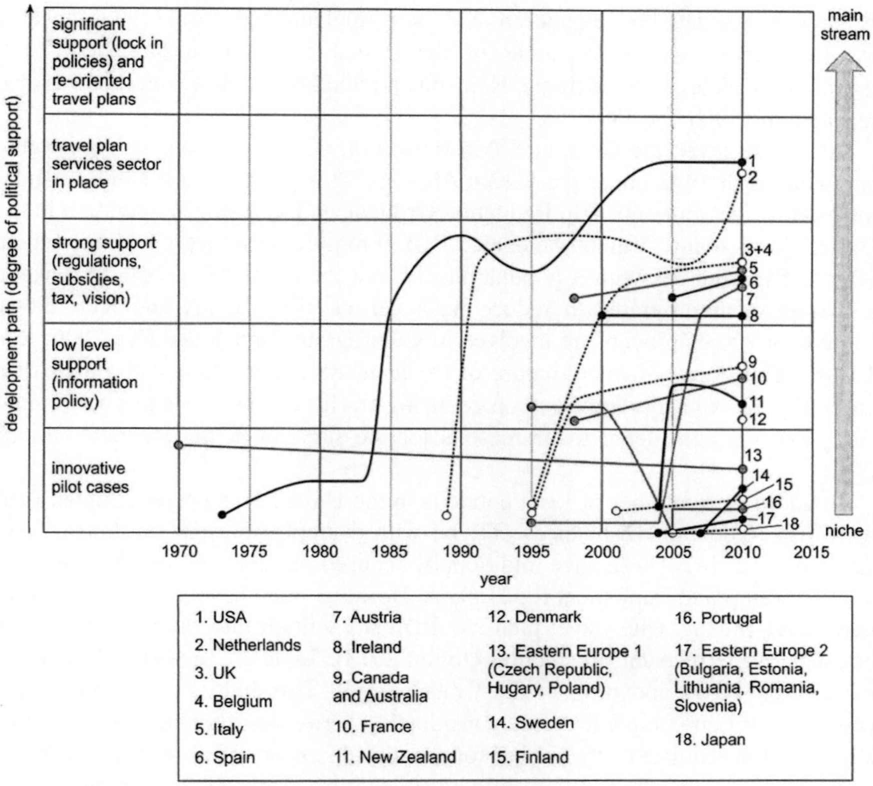

Fig. 2.3 Evolution of travel plans in various countries. *Source* Enoch (2012)

2.4.1 United States: A Strong Regulatory Approach with a Focus on Workplaces

Approaches to travel planning in the United States have largely focused on the use of regulation, with employers as the key target group. Commencing in 1988, Regulation XV in Southern California required employers with 100 or more staff to prepare a travel plan, which was to include the designation and training of an on-site travel plan coordinator (known locally as an employee transportation coordinator) (Giuliano et al. 1991). As part of Regulation XV, an organisation's Average Vehicle Ridership (AVR), defined as the ratio of employees to vehicles arriving at the workplace between 6 am and 10 am, was to achieve specific targets: 1.75 for Central Business Districts (CBDs), 1.5 for other developed urban and suburban areas, and 1.3 for outlying low density areas (Giuliano et al. 1991; Lopez-Aqueres 1993; Orski 1993). Around 9,000 firms employing 3.8 million workers were subject to the regulation (Giuliano et al. 1991). However, the legislation was repealed due to lobbying from businesses in 1996 (Dill 1998; Ferguson 2000).

A similar mandate to Regulation XV was implemented from 1992 onwards in other large metropolitan areas of the United States, including New York, Baltimore, Chicago, Philadelphia, Houston and San Diego. However, this was also repealed in 1996 (Rye 1999b).

At a state level, the Commute Trip Reduction (CTR) program in Washington commenced in 1991 under the ‘Clean Air Act’. This program is the only mandatory state-wide commute Trip Reduction Ordinance (TRO) still in operation in the United States today (Vanoutrive et al. 2010). It requires employers with more than 100 staff in counties with a population of more than 150,000 people to develop and implement programs to reduce peak period trips (Enoch and Rye 2006). More than 700 employers are involved in the program (Enoch and Rye 2006; Rye 1999b). The ‘locally-owned’ nature of the mandate (passed at a state rather than federal level) and the emphasis on reducing traffic congestion and energy consumption, are considered to be reasons for the survival of the mandate (Enoch 2012).

In addition, a number of local councils in the United States have adopted their own Trip Reduction Ordinances (TROs) with examples present in almost every state (ABC 2014). These have traditionally focused on requiring existing employers to develop and implement travel plans. However, they have also been used for new developments with travel plan requirements written into the municipal code or plans of the relevant jurisdiction (Jollon 2013). Table 2.2 provides a summary of selected TROs operating in the United States. Thresholds, requirements and penalties for non-compliance vary considerably across jurisdictions. For example, Bloomington requires a travel plan for any developments of more than 1,000 ft^2, while Minneapolis adopts a threshold of 100,000 ft^2. Some TROs prescribe a specific set of travel plan measures that should be implemented while others are more open ended by simply requiring the preparation of a travel plan. In some jurisdictions, financial penalties apply for not complying with the TRO, yet other jurisdictions do not adopt any penalties for non-compliance.

2.4.2 United Kingdom: A Large Focus on Requiring Travel Plans for New Developments

In 1998, only 3 % of local councils in the United Kingdom had implemented a travel plan on a permanent basis, with a further 4 % implementing one on a trial basis (Bradshaw et al. 1998). The level of travel plan take-up substantially increased in future years. A survey conducted in 2000 by Steer Davies Gleave (2001) found that 24 % of councils had a travel plan in place, while a further 45 % were developing one. In addition, a number of businesses, hospitals and higher education establishments were also involved in travel planning. The survey also revealed that 58 % of councils had required travel plans for new developments as part of the planning approvals process. In a subsequent survey of councils by

Table 2.2 Characteristics of selected trip reduction ordinances (TROs) in the United States

Location	Thresholds	Requirements	Penalties for non-compliance
Boston, Massachusetts	Any development with more than 50,000 ft^2 or 15+ dwelling units	Recommended list of travel plan measures	None specified
Bloomington, Minnesota	Any development with more than 1,000 ft^2 or 350 parking spaces	Travel plan with outline of expenditures over 3 years	$50 per parking space (no timeframe specified)
Boulder, Colorado	Any development adding 20+ peak trips for residential or 100+ peak trips for commercial	Choice of one of three travel plans: transit oriented, parking management, other	None enforced
Cambridge, Massachusetts	Non-residential development with more than 5 parking spaces	Specific set of travel plan measures, including on-site travel plan coordinator	$10 per parking space per day
Durham, North Carolina	Any business with more than 100 employees	List of suggested travel plan measures to choose from	$100/day up to maximum of $1,000
Fairfax, Virginia	All new developments	Preparation of travel plan	None specified
Minneapolis, Minnesota	Any development with more than 100,000 ft^2	Preparation of travel plan	None being applied
Montgomery, Maryland	Any business with more than 25+ employees	Travel plan plan, including employee transport coordinator	None specified
Pasadena, California	Any development with more than 25,000 ft^2	Preparation of travel plan with list of suggested measures	Permit withheld, fines
Pima County, Arizona	Any business with more than 100 employees	Annual survey and travel plan with specific measures required	Up to $250/day
Portland, Oregon	Businesses with more than 100 employees in portland air quality maintenance area	List of suggested travel plan measures	Revised plan to be submitted with proof of good faith effort
Richmond, California	Any business with more than 10 employees	Must provide a pre-tax benefit, transit subsidy or shuttle service	$200–$600 (no timeframe specified)
Rockville, Maryland	Development adding 30+ trips in peak period	Payment of transport improvement fee and/ or preparation of travel plan	None specified
San Francisco, California	Any business with more than 20 employees	Must provide a pre-tax benefit, transit subsidy or shuttle service	$100/year

(continued)

Table 2.2 (continued)

Location	Thresholds	Requirements	Penalties for non-compliance
Santa Monica, California	Any business with 10+ employees	Preparation of travel plan containing specific measures	$5/day per employee
Seattle, Washington	Any business with 100 full-time employees	Two travel plan measures from a specific list	$250/day
State of Massachusetts	Businesses with 1,000+ employees/students	List of specific travel plan measures	None

Source Author's synthesis of the literature based on ABC (2014), Jollon (2013) and Stewart (1994)

Addison & Associates (2008), a threefold increase had been found in the number of travel plans required through the planning process between 2001 and 2006. In addition, 65 % of councils had required over 10 travel plans since 2001, with three councils requiring over 200 during that timeframe.

Requirements for travel plans at new developments are well supported through a national planning policy in the United Kingdom (Rye et al. 2011b), with development 'thresholds' used to specify when a travel plan is required. Examples adopted by Transport for London (2011b) include developments with at least 80 residential units, 1,000 m^2 of retail space or 2,500 m^2 of office. However, it is also noted that thresholds can vary considerably, even between local councils that are located within the same region (Addison & Associates 2008).

Morris et al. (2009) refer to a number of travel plans prepared for new residential developments in the United Kingdom. Examples include the Beddington Zero Energy Development (BedZED) in London, Poole Quarter in Dorset, and various residential sites in Nottingham. However, he notes that 'there is much policy guidance on developing successful workplace travel plans but still relatively little information is currently available regarding the potential for applying similar principles to residential sites' (Morris et al. 2009, p. 25). Furthermore, the UK Department for Transport (2005, p. 3) recognise that there is a 'limited scale of experience nationally of developing residential travel plans' and that 'it should be recognised that emerging good practice is at an embryonic stage'. Moreover, Addison & Associates (2008, p. 12) state that 'there is no indication of the number of residential travel plans that have been drawn up or implemented' and that their research was 'unable to ascertain numbers as the majority of councils do not monitor their travel plan activities'.

Other travel planning activity reported in the United Kingdom includes:

- **Schools**: by 2003, around 3,100 school travel plans in the United Kingdom had been implemented (Cairns et al. 2004). However, Smith (2010) report that this had increased to over 10,000 by 2006, representing around 40 % of the potential 24,000 school travel plans.

- **Hospitals**: Khandokar et al. (2013) found that by 2011, 115 acute trusts out of a total of 170 National Health Service (NHS) acute trusts (68 %) had a travel plan in place.
- **Railway stations**: ATOC (2013) report that there are currently more than 70 travel plans for train stations in place, some of which have been required through the planning process.

2.4.3 Australia: A Focus on Voluntary Adoption by Workplaces and Schools

In contrast to the United States and United Kingdom, Australia has taken more of a voluntary approach to travel planning, with a focus on pre-existing sites such as workplaces and schools.

An expression of interest process is used in Western Australia to seek employers wishing to develop a travel plan (Thom 2009). Workplaces are selected according to their level of organisational commitment and geographic location (e.g. proximity to public transport). A similar approach is used in South Australia whereby interested workplaces need to demonstrate a level of commitment to resourcing their travel plan (Halling and Mayes 2011). In the state of Victoria, workplace travel plans have been prepared on behalf of employers (DeGruyter et al. 2005; Meiklejohn and Wake 2007), however these were found to be less successful as the workplace community was removed from actively participating in the process of developing the travel plan, thereby reducing the level of ownership in the concept (Howlett and Watson 2010).

Moghtaderi et al. (2012) report on the scale of voluntary school travel planning practice in Australia. They describe activity in the states of Victoria (more than 30 schools since 2002), New South Wales (15 schools), Western Australia (30–60 schools per year since 1998, with 160 schools involved by 2006), South Australia (190 schools by 2010) and Queensland (117 schools since 2004).

Voluntary adoption of travel plans is also evident across Australia at various hospitals (McFadden et al. 2006; Petrunoff et al. 2013), shopping centres (Woodruff and Hui 2010) and universities (Cooper and Meiklejohn 2003; Curtis and Holling 2004).

In the context of new developments, no national or state policy is in place in Australia that supports the requirement for travel plans. Requirements for travel plans are generally only specified by some councils through local planning policies. In the state of Victoria, only 4 out of the 79 councils (5 %) explicitly include a requirement for travel plans in their local planning policy (Department of Transport, Planning and Local Infrastructure 2014b). Despite this minor representation, other councils are not precluded from imposing the requirement where it is considered appropriate. In the state of New South Wales, travel plans have been required for new developments on a mostly ad hoc basis; examples include a mixed use development in Rouse Hill (Wiblin et al. 2012) and the relocation of Optus, a major telecommunications employer (NSW Government 2011). More recently, councils in Western Australia are beginning to consider requirements for travel plans for new developments, although this is still in the development stages (RAC 2014). While the literature cites these few examples, **there is no understanding of the scale and associated characteristics of travel planning practice for new developments in Australia**.

2.4.4 Other Countries

Examples of travel planning activity occurring in other countries include:

- **New Zealand**: 178 schools, 31 workplaces and 2 tertiary institutions have been involved in travel planning since 2008 (Sullivan and Percy 2008). A limited number of councils (Waitakere City Council, Rodney District Council and Hamilton City Council) require travel plans for new developments although no national policy exists to support this process (Baker 2007).
- **Canada**: more than 100 schools have been involved in travel planning since 2010 (Mammen et al. 2014). In addition, the City of Mississauga in Ontario requires a set of specific travel plan measures to be implemented at new residential developments (Mele 2013), while the City of Waterloo in Ontario provides a voluntary TDM checklist for non-residential developments to support proposed reductions in car parking provision (Hill 2013).
- **Italy**: in 1998, the government mandated that employers with over 300 staff nominate a mobility manager to reduce the impact of commuting trips through a travel plan (Enoch and Potter 2003). However, no quantitative targets are set and there are no penalties for employers that do not comply with the requirement (Potter and Enoch 2007).
- **Belgium**: the Brussels capital region requires a travel plan for every employer with at least 200 employees (Vanoutrive et al. 2010).

In addition, there are various examples of car-free housing developments located in Germany, Austria, the Netherlands and Denmark (Melia et al. 2013; Wright 2005). While these do not typically have travel plans in place, they often include an on-site car sharing service (Wright 2005).

In summary, travel plans have been adopted across many countries including the United States, United Kingdom, other parts of Europe, Australia, New Zealand and Canada. Most of these countries have required travel plans for new developments in some form, yet various issues have arisen in doing so which are discussed in the next section.

2.5 Issues with Requiring Travel Plans for New Developments

Perhaps as a reflection of the relative novelty of requiring travel plans for new developments, a number of issues have been experienced in their development, implementation and monitoring. Table 2.3 provides a synthesis of the issues cited by the literature.

Travel plans of varying quality have been submitted and subsequently approved, particularly when they are solely prepared to seek planning approval. While clear guidance may help to address the problem, the United Kingdom have still experienced instances of poor travel plan quality despite the availability of various guidelines, standards and tools (British Standards Institution 2008; Department for Transport 2009; Transport for London 2011a, b; Atkins 2002). While the literature suggests there is an issue relating to travel plan quality (Addison & Associates 2008; Enoch and Ison 2008; Melia 2009; Wynne 2013), **no formal assessment of the quality of travel plans prepared for new residential developments has been undertaken to identify their relative merits and potential areas for improvement**.

Table 2.3 Key issues in requiring travel plans for new developments

Stage	Key issues
Developing the travel plan	• Developers paying 'lip-service' to the concept, particularly when the travel plan is only prepared to seek planning approval • Lack of travel planning guidance specifically for new developments in some jurisdictions • Varying quality of travel plans being submitted and approved, particularly where planning assessment officers lack sufficient knowledge or experience with travel plans • Travel plan considered too late in the land use planning process • Nature of proposed development is sometimes unknown, despite this information being vital to informing the objectives, targets and measures in a travel plan
Implementing the travel plan	• General lack of implementation of travel plan measures and associated follow-up • Lack of suitable handover arrangements from the developer to tenant or property manager • Inconsistency between the objectives of the travel plan and motivations of those responsible for implementing the travel plan • Uncertainty over roles and responsibilities • Lack of ownership of the travel plan
Monitoring the travel plan	• General lack of monitoring, leading to a lack of evidence of travel plan effectiveness • Insufficient resources within local government to undertake effective enforcement • Uncertainty in use of legal mechanisms for enforcing travel plans • Uncertainty over roles and responsibilities

Source Author's synthesis of the literature based on Addison & Associates (2008), Enoch and Ison (2008), Hendricks (2008), Llewellyn et al. (2014), Roby (2010c), Rye et al. (2011a) and Wynne (2013)

A lack of implementation is consistently highlighted as a key issue associated with travel plans for new developments. For example, in a survey of 69 councils in the United Kingdom, Addison & Associates (2008, p. 66) found that 'less than half of the travel plans they had required for new developments had been implemented'. Roby (2010b, p. 42) states that the 'lack of, or poor implementation of a travel plan as part of the planning process is a problem with this method of securing travel plans.'

In the context of new residential developments, it is worth noting the inherent difficulty associated with implementation. Firstly, in contrast to the more traditional workplace or school travel plan, residential travel plans are based on the trip origin and therefore need to cater not only for a range of trip destinations, but also for a range of trip purposes (Morris et al. 2009). Secondly, the need to establish an ongoing management structure to deliver a residential travel plan presents challenges as there is often a weak relationship between the residential provider and the residents themselves (Enoch 2012). Despite these issues, **there has been no research undertaken to sufficiently explore implementation in the context of travel plans for new residential developments**.

A lack of monitoring and enforcement of travel plans for new developments is also highlighted by the literature. The main reason for this is a lack of resources within councils:

> Some authorities stated that once approved the implementation of these travel plans was neither monitored nor enforced. Resourcing (or the lack of it) of the monitoring, penalties, sanction and incentives processes was seen by many authorities as a reason for not including them within travel plans as they have no resources to follow this through (Addison & Associates 2008, p. 71).

> Even after a physical development is completed, promises, both verbal and written, about program implementation and what will or will not be permitted to take place in future, are often forgotten or not enforced…Agreements to engage in TDM are often not followed over time (Seggerman and Hendricks 2005, pp. 61–2).

The UK Department for Transport (2009) suggest that if local councils establish a fee or secure an agreed sum for travel plan monitoring, this can be pooled over a number of councils to provide a cost effective solution for all parties involved. This has recently been put in practice with WestTrans, a partnership of six London councils, managing the process associated with monitoring development related travel plans in West London. A monitoring officer position is funded on a full-time basis collectively across the councils, representing the first group of councils in England to take this approach to travel plan monitoring (Khagram 2013a).

2.6 Evaluating the Effectiveness of Travel Plans

A range of methods can be used to evaluate the effectiveness of travel plans, depending on the size and location of the site, the type of travel plan, and its objectives. Table 2.4 presents a synthesis from the literature of common methods used.

Table 2.4 Common methods used to evaluate the effectiveness of travel plans

No.	Method	Strengths	Limitations
1	Assessment of take-up rates of travel plan measures	Determines relative success of different measures; helps in targeting future travel plan measures	Unable to measure outcomes such as changes in car use
2	Assessment of travel plan document quality	Provides proxy for effectiveness; can help to identify areas for improvement	Unable to measure outcomes such as changes in car use; poor quality travel plan could still result in a successful outcome
3	Assessment of level of travel plan implementation	Identifies extent of implementation which may help in explaining outcomes	Unable to measure outcomes such as changes in car use; full implementation does not guarantee a successful outcome
4	Trend assessment of public transport patronage data	Indicates changes in public transport use over time	Dependent on data availability; difficult to control for external factors (e.g. service changes, fuel prices, population growth)
5	Travel survey questionnaire	Can help to determine transport modal split, awareness of travel plan measures and other indicators	Response rates vary; needs to be conducted regularly or compared to secondary data or control sites
6	Hands-up survey	Provides a quick and easy method for determining transport modal split	Typically limited to classroom/school environments; concerns over reliability
7	Focus groups	Can explore impacts of the travel plan in depth with users of the site	Limited to small groups and generally only provides qualitative information; can be labour intensive
8	Vehicle counts	Provides an independent measure of car use at the site	Can be labour intensive; needs to be conducted regularly or compared to secondary data or control sites to measure effectiveness
9	Multi-modal counts	Provides an independent measure of transport modal split	Can be labour intensive; needs to be conducted regularly or compared to secondary data or control sites to measure effectiveness
10	Car parking counts	Determines utilisation of car parking at site and can help to inform future travel plan measures	Needs to be conducted regularly or compared to control sites to measure effectiveness
11	Bicycle parking counts	Determines utilisation of bicycle parking at site and can help to inform future travel plan measures	Needs to be conducted regularly or compared to other sites to measure effectiveness

Source Author's synthesis of the literature based on Addison (2002), Ampt et al. (2009), Halling and Mayes (2011), Higgins (1996), Hinckson and Badland (2011), Stewart (1994), Transport for London (2008), Travel Plan Services Ltd (2013), TRICS (undated), Wake (2012), Wake et al. (2010), Wiblin (2010) and Wiblin et al. (2012)

Methods 1–3 (assessments of travel plan take-up, quality and implementation) are unable to measure outcomes such as changes in car use, but nonetheless provide useful indicators to support the overall assessment of travel plan effectiveness (Wake et al. 2010). Method 4 (trend assessment of public transport patronage data) is dependent on data availability and may pose difficulties in controlling for external factors such as population growth.

Method 5 (travel survey questionnaire) is the most common technique used to evaluate travel plans, particularly at pre-existing sites (Steer Davies Gleave 2001). A 'before' travel survey is typically conducted to provide a baseline estimate of travel patterns and to inform the development of the travel plan. An 'after' survey, usually conducted annually thereafter, is then used to assess any changes in travel behaviour that may have occurred (Ampt et al. 2009). However, in the context of new developments, baseline travel patterns are generally not available as the development is not usually occupied or even built. While it is possible to conduct a before survey shortly after occupation of the development, travel plan measures may have already been introduced at the site making it difficult to measure their impact (Stewart 1994).

Method 6 (hands-up survey) is typically limited to schools and provides a relatively simple method to estimate the transport modal split. The reliability of this method is often questioned due to students being influenced by others in choosing 'popular' transport modes. However, a pilot study undertaken by Hinckson and Badland (2011, p. 370) showed '100 % agreement between children's and parents' responses when parents were telephoned the same afternoon'. Method 7 (focus groups) is commonly adopted in conjunction with other methods, but is more suited to informing the development of the travel plan at pre-existing sites, rather than in determining its effectiveness. This is due to the small number of participants that are usually involved and the qualitative nature of the technique.

Methods 8–9 (vehicle and multi-modal counts) provide an independent measure of trip generation and the transport modal split at the site. At pre-existing sites, the measurements need to be undertaken before and after the travel plan is introduced in order to assess its impact. However, at new developments, count data is often compared to secondary data sources, such as published vehicle trip generation rates or regional travel survey data. However, data comparability issues can arise in terms of the geographic location, target population and data collection period. In these instances, control sites can be adopted although these need to be chosen carefully to ensure they exhibit similar characteristics to sites with travel plans. Methods 10–11 (car and bicycle parking counts) can assist in evaluating the effectiveness of travel plans, but again need to be compared to 'before' measurements in the case of pre-existing sites, or to secondary data sources or control sites in the case of new developments.

Given that each method has both strengths and limitations, a mixed methods approach is recommended so that reliance is not placed on a single evaluation measure (Wake et al. 2010).

Evaluations of travel plans in reducing car use have been widely reported. Table 2.5 provides a summary of evaluations conducted in the United Kingdom,

Table 2.5 Evaluations of workplace and school travel plans in reducing car use

Country	Key findings
United Kingdom	• Average reduction in car use of 18 % across 20 workplaces (Cairns et al. 2004) • Average reduction in car use of 11 percentage points (or 17 %) across 41 workplaces (Bamberg and Moser 2007) • Average reduction in car use of 23 % across 28 schools, with two schools achieving a reduction of more than 50 % (Cairns and Newson 2006)
United States	• Average reduction in car use of 15 % across 49 employer sites (TCRP 1994 cited in Cairns et al. 2010) • Average reduction in car use of 6.3 percentage points (from 73.5 to 67.2 %) across 5,000 employers in Southern California subject to Regulation XV (Young and Luo 1995) • Average reduction in car use of 5.5 percentage points (from 80.1 to 74.6 %) across 31 employers in Chicago (Pagano and Verdin 1997)
The Netherlands	• Average reduction in vehicle kilometres of 8 % for travel plans with 'basic' measures and 20 % for travel plans with 'luxury' measures, across 40 employers (Ligtermoet 1998 cited in Cairns et al. 2010)
Australia	• Average reduction in car use of 5 percentage points across 12 employers in Western Australia, with a benefit cost ratio of 4.5–1 (Marsden Jacob Associates 2011) • Average reductions in car use ranging from 8 to 35 % across more than 200 schools across different states (Moghtaderi et al. 2012) • Reductions in car use of up to 15 percentage points across eight primary schools, three workplaces and one university in Victoria (Department of Transport undated-a, undated-b, undated-c)
New Zealand	• Increase in active travel of 5.9 percentage points (from 34.9 to 40.8 %) across 33 schools in Auckland (Hinckson and Badland 2011) • Average reduction in car use of 3.4 percentage points across 68 primary and secondary schools in Auckland (Hinckson et al. 2009)
Canada	• No significant change in active travel across 53 schools; at a disaggregate level, changes in active travel ranged from −26 to +23 % in the AM peak period and from −24 to +15 % in the PM peak period (Mammen et al. 2013) • 17 % of parents across 103 schools reported less driving (Mammen et al. 2014)

Source Author's synthesis of the literature based on citations within the table

United States, the Netherlands, Australia, New Zealand and Canada. The evaluations incorporate both pre-existing sites and new developments but are focused solely on workplaces and schools. Results vary considerably but are generally in the order of a reduction in car use of 10–15 %. Consistent with other research concerning the effectiveness of travel plans (Cairns et al. 2008; Miller 1995; Rye 1999a), Enoch and Rye (2006) state that reductions in car use of different travel plans are generally minimal when including information-only measures, 5 % for mainly carpooling-related measures, 8–10 % for travel plans that incorporate financial incentives to using non-car modes, and 15 % or more for travel plans that include financial disincentives to car use (e.g. car parking charges).

Long-term monitoring of travel plans has been limited to date. Figure 2.4 presents a synthesis of results for workplace travel plans from the United States, United Kingdom, Australia and New Zealand. A continual reduction in car use

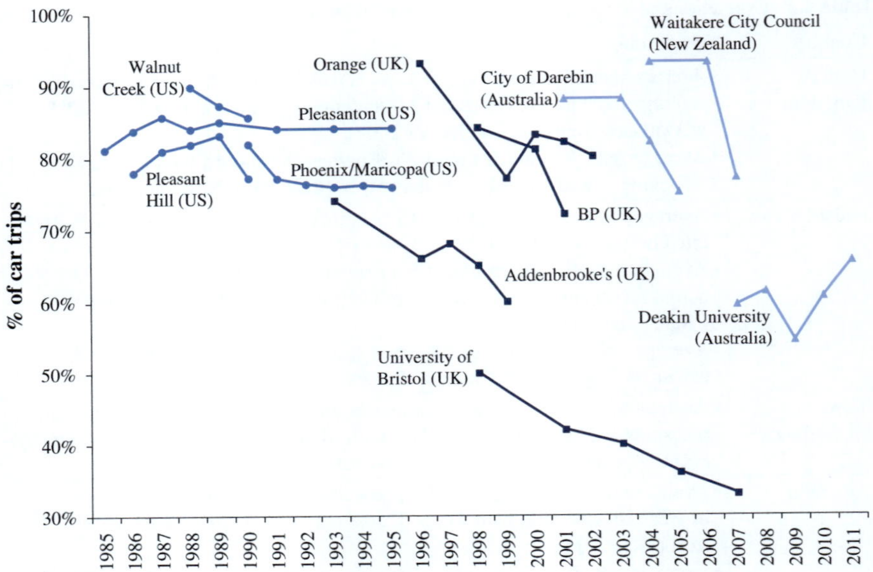

Fig. 2.4 Results from long-term monitoring of workplace travel plans. *Source* Author's synthesis of the literature based on Baker (2007), Brockman and Fox (2011), Cairns et al. (2002), Dill (1998), Hancock and Nuttman (2013) and Myers (2005)

(compared to baseline levels) is evident in most cases, although some sites did experience an increase in car use at times during their evaluation period. There may also be cases of less successful results that have simply not been published, possibly because researchers choose not to publish them or have difficulty in publishing negative results (Bamberg and Moser 2007; Richter et al. 2011). Cairns et al. (2002, p. 85) note that the duration of a travel plan is not critical to its success '… because the measures contained in the travel plan appear to be more important than simply how long the organisation has been attempting to address the problem'.

Evaluations of travel plans for new residential developments have been both limited in number and scope. Table 2.6 presents a summary of evaluations conducted in the United States, United Kingdom and Australia. Reductions in car use were found in most cases; however these were mostly based on comparisons to secondary data sources (e.g. census and regional travel survey data) that were inconsistent with the evaluations conducted at the travel plan sites. For example, the evaluations were undertaken during different years to when the secondary data was collected. Also, the secondary data generally covered all housing types, yet the travel plan sites were generally not representative of all housing types. This may therefore lead to differences in car parking provision and the socio-demographic characteristics of residents, which will ultimately affect the comparison of travel patterns.

Table 2.6 Summary of evaluations of travel plans for new residential developments

Sites	Key findings	Limitations
16 residential developments Arlington, United States (Arlington County Commuter Services 2013)	Vehicle trip generation up to 60 % lower than published rates 51 % of residents drove alone to work, compared to average of 54 % for Arlington and 64 % for wider region (based on regional travel survey data)	Published vehicle trip generation rates not comparable to sites in terms of geographical location and data collection periods Sites not representative of all housing as per regional travel survey data, particularly in terms of car parking provision
8 residential developments London, United Kingdom (WSP 2014)	10 % less car use than average for surrounding area (based on census data)	Data not comparable in terms of survey periods: observed data collected in 2013; census conducted in 2011 Sites not representative of all housing as per census, particularly in terms of car parking provision
Beddington Zero Energy Development (BedZED) South London, United Kingdom (BioRegional 2009)	Fossil fuel car mileage 65 % less than national average 17 % of residents' work trips were by car, compared to 49 % for surrounding area	Comparisons to regional averages may not be comparable in terms of data collection periods Issue of residential self-selection not addressed given nature of zero energy development
Poole Quarter Poole, United Kingdom (Department for Transport 2005; Melia 2009)	43 % reduced car use compared to previous place of residence Proportion of households with no car (19 %) similar to regional average (18 %)	Impacts of travel plan not adequately assessed as reductions in car use were largely due to changes in residential location
Newcastle Great Park Newcastle Upon Tyne, United Kingdom (Travel Plan Services Ltd 2013)	Car mode share target of 77 % met (actual was 71 %) Public transport target of 18 % not met (actual was 13 %) 62 % increase in bus patronage since 2010–12	No comparisons made to surrounding area; basis and justification for targets is unclear
Queen Elizabeth Park Guildford, United Kingdom (Department for Transport 2005)	Car mode share target of 62 % not met (actual was 80 %); yet only 37 % of trips made by car as single occupant	No comparisons made to surrounding area; basis and justification for targets is unclear
Rouse Hill New South Wales, Australia (Wiblin et al. 2012)	Car ownership of 1.6 veh/household compared to 2.0 veh/household for control suburbs; greater use of bus, walking and cycling compared to control suburbs	Control suburbs not directly comparable as they were more established and had less accessibility to town centre facilities

Source Author's synthesis of the literature based on citations within the table

The use of control sites with similar characteristics, and with data collected at the same time as the treatment (travel plan) sites, can better account for external factors and provide a more accurate indication of travel plan effectiveness. This approach was identified by Arlington County Commuter Services (2013) as a future research need in their evaluation of travel plans at 16 residential developments in Arlington, United States. However, of the evaluations presented in Table 2.6, only one (Rouse Hill, New South Wales, Australia) used a control group. Yet even in this case, the treatment (travel plan) group was considered to have better access to town centre facilities than the control group (Wiblin et al. 2012), thereby potentially impacting upon the level of car use.

While there are other examples of travel plans implemented at new residential developments (Department for Transport 2005) in addition to those shown in Table 2.6, evaluations of these have either not been published or undertaken at all. Addison & Associates (2008, p. 20) notes:

> …there has been no overarching research study to evaluate the effects of residential travel plans on daily car trip rates per unit or modal share at new developments. Many of the residential travel plans have not been monitored or have yet to be fully operational.

More recent work has confirmed this is still the case. Khagram (2013b, 2014) reports on a monitoring program focused on development-related travel plans in London. As of Spring 2013, only 12 out of 242 sites had completed their monitoring program with no results available for residential sites (Khagram 2013b).

Furthermore, no evaluations of travel plans for new residential developments have yet accounted for self-selection bias effects. For example, residents who choose to live at a development with a travel plan may already be more disposed to using sustainable forms of transport. Therefore, any differences in travel behaviour that are observed when comparing to secondary data or control sites could be the result of residential self-selection and not the travel plan itself.

Hence, little research has been undertaken to appropriately quantify the effectiveness of travel plans in reducing car use at new residential developments, with no studies accounting for self-selection effects.

2.7 Success Factors for Travel Plans

A wide range of success factors for travel plans have been cited in the literature. A full list is provided in Appendix B. Key factors that are consistently reported include:

- Building ownership and engagement in the travel planning process (Howlett and Watson 2010)
- Securing senior management support (Baudains 2003)
- Having an enthusiastic and dedicated travel plan coordinator (Van Malderen et al. 2013)
- Implementing a set of comprehensive travel plan measures (Ison and Rye 2008)

- Incorporating constraints on car parking (Cairns et al. 2010)
- Having a supportive policy framework in place (Addison & Associates 2008).

The need to build ownership and engagement in the travel planning process is generally common to all travel plans. However, for new developments, Harrison (2003, p. 400) cautions that:

> …travel plans are increasingly being drafted for applicants by consultants. While this is welcome, in that a body of knowledge and expertise is being built up by specialists, it carries the risk that no one in the applicant's organisation has any particular personal commitment to making the plan a success. Indeed the individual who may feel most committed to the travel plan, having drafted and negotiated it, may be the consultant who will have no further connection with the site once planning permission has been granted.

Where a developer is involved in preparing the travel plan, Yeates and Enoch (2012, p. 13) recognise that 'ownership of a travel plan in the long term is a difficult issue, particularly where financial conditions are involved in regard to monitoring'. The UK Department for Transport (2007, p. 13) advise that the success of travel plans at new residential developments 'depends on ensuring that ownership for the plan ultimately rests with the residents who recognise the benefits and are aware that the plans are in their best interest'.

Securing senior management support is relevant for workplaces, particularly where funding may be required to implement initiatives or support for more controversial measures is required (Baudains 2003; Cairns et al. 2010; Rye 1997). However, it is also applicable to schools where support from the school principal is critical to the success of the travel plan (Newson et al. 2010).

The importance of having an enthusiastic and dedicated travel plan coordinator is well documented (Hendricks and Georggi 2007; Van Malderen et al. 2013). Hendricks (2005) investigated the impact of travel plan coordinator styles on the effectiveness of travel plans and found that the more successful ones tended to have travel plan coordinators with 'influencing' and 'steady' work styles. Rye (1997) found that a number of personal abilities are particularly important if a travel plan coordinator is to be effective. These include, among others, negotiating abilities, resilience, and an ability to deal tactfully with people.

Implementing a set of comprehensive travel plan measures that work together as an integrated package is considered to be an important component of successful travel plans, part of which involves tailoring the choice of measures to the needs of the site and its users (Cairns et al. 2010; Fraser and Addison 2002; Ison and Rye 2008). However, Cairns et al. (2010, p. 492) found that the amount of money spent on travel plans 'did not relate directly to the degree of change that had been achieved, or the overall 'end' level of car use … the appropriateness of the measures and overall strategy appeared to be more important to travel plan effectiveness'. Furthermore, Orski (1993, p. 162) found that 'large expenditures do not always ensure program success, and lesser expenditures can sometimes be as effective, if appropriately targeted'.

Car parking constraints have also been cited widely in the literature as a key success factor for travel plans. Cairns et al. (2010) report the results of around 20 case studies in the United Kingdom which showed that workplaces that had addressed parking achieved more than double the reduction in car use of those that had not. Research undertaken into employer transport benefits in the United States revealed a strong relationship between the provision of free car parking and driving to work, with free car parking resulting in an increase of 20 % points for driving alone to work in the Washington DC region (Hamre and Buehler 2014).

The need for a supportive policy framework is highly relevant to travel plans for new developments. Through a survey of 69 local councils in the United Kingdom, Addison & Associates (2008) found that a strong policy context was the most often cited 'assisting' factor in securing travel plans. The UK Department for Transport (2005, p. 57) also note:

> A robust policy framework is important in supporting negotiations to secure residential travel plans…The more comprehensive, integrated and explicit the authority's policy framework is in relation to sustainable spatial planning and transport requirements and the role of travel plans, the more easily a requirement for a residential travel plan can be justified.

Despite the range of success factors for travel plans that have been cited, efforts to quantify their relative effectiveness have been limited. Winters et al. (2005) developed a model to predict the change in vehicle trips at workplaces based on site characteristics and the specific travel plan measures introduced. However, despite the use of data from several thousand employer sites, the best model achieved a relatively low level of agreement with actual results (the best model explained only 21.5 % of the variability in the data; an R^2 value of 0.215). The authors note that the 'overall poor accuracy for all the models can be attributed more to the complexity of this problem involving unpredictable human behaviour' (Winters et al. 2005, p. 204). They also note 'the effect of the three Es—empowerment, experience, and enthusiasm—of the employee transportation coordinator' (Winters et al. 2005, p. 206), factors that can be difficult to incorporate within a quantitative model.

Similarly, Orski (1993, p. 162) found that 'the effectiveness of (travel plan) programs depends to a large extent on intangible, difficult-to-quantify factors: the commitment of senior management, the aggressiveness with which the program is promoted, and the status and visibility of the Employee Transportation Coordinator'. Cairns et al. (2004, p. 25) also found that 'very few generalisations that could be made' when attempting to identify why some travel plans are more successful than others.

2.8 Conclusion

The aim of this chapter is to provide a review of the literature on travel plans and their application to new developments. In doing so, it has covered the travel planning process, actors, geographical coverage and scope, issues, effectiveness and key success factors. Research gaps and opportunities have been identified from the literature review and are summarised in Table 2.7.

Table 2.7 Research gaps and opportunities

Section	Research gaps	Research opportunities
2.3 Actors involved in the travel planning process	No research has specifically explored the perspectives of the different actors involved in travel planning for new residential developments	Develop an appreciation for the perspectives of actors involved in travel planning for new residential developments, particularly aspects relating to implementation (see Chap. 6)
2.4 Geographical coverage and scope	There is no understanding of the scale and associated characteristics of travel planning practice for new developments in Australia	Examine the scale and associated characteristics of travel planning practice for new urban developments using a case study from the state of Victoria (see Chap. 5)
2.5 Requiring travel plans for new developments	No formal assessment of the quality of travel plans prepared for new residential developments has been undertaken	Undertake a quantitative assessment of the quality of travel plans prepared for new residential developments to help identify their relative merits and potential areas for improvement (see Chap. 7)
	No research has been undertaken to sufficiently explore implementation in the context of travel plans for new residential developments	Explore the implementation process associated with travel plans for new residential developments to identify opportunities to enhance effectiveness (see Chap.9)
2.6 Evaluating the effectiveness of travel plans	Little research has been undertaken to appropriately quantify the effectiveness of travel plans in reducing car use at new residential developments, with no studies accounting for self-selection effects	Using a case study approach, evaluate the impacts of travel plans for new residential developments including self-selection effects to understand their effectiveness in reducing car use (see Chap. 8)

The literature is relatively silent on the scale of travel planning practice for new developments in Australia, apart from a few specific examples (Department of Transport, Planning and Local Infrastructure 2014b; NSW Government 2011; RAC 2014; Wiblin et al. 2012). This thesis will help to close this gap by assessing the scale and associated characteristics of travel plans for new urban developments, using a case study from the state of Victoria (see Chap. 5).

While previous research has explored the perspectives of actors involved in travel planning (Davison et al. 2010; Enoch and Ison 2008; Rye et al. 2011a; Yeates and Enoch 2012), none has been undertaken specifically within the context of new residential developments. This is particularly relevant given the different implementation challenges associated with residential sites. This thesis explores these issues from the perspectives of relevant actors involved in the process (see Chap. 6). Related to this is the lack of research that has sufficiently explored implementation in the context of travel plans for new residential developments. This thesis will address these research gaps and in doing so will identify opportunities to enhance the effectiveness of travel plans for new residential developments (see Chap. 9).

Moreover, the effectiveness of travel plans for new residential developments is poorly understood. (Addison & Associates 2008; Morris et al. 2009). Using a case study approach, this thesis will address this issue by providing an understanding of their effectiveness in reducing car use at new residential developments, including the quantification of any self-selection bias effects (see Chap. 8). A quantitative assessment of the quality of travel plans prepared for new residential developments will also be undertaken to help identify their relative merits and potential areas for improvement (see Chap. 7).

Prior to outlining the specific methodology for addressing each of the research gaps, this literature review continues in the next chapter, but with a shift towards the theoretical foundations of the research. This includes coverage of both implementation theory and planning enforcement theory.

References

ABC. (2014). *Establishing an effective commute trip reduction policy in Massachusetts*. Massachusetts, US: A Better City (ABC).

ACT Canada & Noxon Associates Limited. (2010). *Workplace travel plans: Guidance for Canadian employers*, Transport Canada.

Addison, L. (2002). *Using the planning process to secure sustainable transport*. Paper presented to European Transport Conference, Homerton College, Cambridge, UK.

Addison & Associates. (2008). *Delivering travel plans through the planning process—Research report*. London, UK: Department for Transport and Communities and Local Government.

Ampt, E. S., Richardson, A. J., & Wake, D. (2009). *Simple and suited: Guidelines for workplace travel surveys*. Paper presented to 32nd Australasian Transport Research Forum (ATRF), Auckland, New Zealand.

Arlington County Commuter Services. (2013). *Residential building transportation performance monitoring study*, Arlington, Virginia, US.

Atkins, W. S. (2002). *Workplace travel plan evaluation tool v2.7*. Department for Transport. Retrieved September 25, 2014, from http://www.imsaho.com/miscellaneous/travel_plan_evaluation_tool.asp.

ATOC. (2013). *Guidance on the implementation of station travel plans*. UK: Association of Train Operating Companies.

Baker, L. (2007). *Workplace travel plans—A new rule of the game*. Paper presented to New Zealand Planning Institute and EAROPH Conference, Palmerston North, New Zealand.

Bamberg, S., & Moser, G. (2007). Why are work travel plans effective? Comparing conclusions from narrative and meta-analytical research synthesis. *Transportation, 34*, 647–666.

Baudains, C. (2003). *Environmental education in the workplace: Inducing voluntary travel behaviour change to decrease single occupant vehicle trips by commuters into the Perth CBD*. PhD Thesis, Murdoch University.

BioRegional. (2009). *BedZED seven years on: The impact of the UK's best known eco-village and its residents*, UK.

Bradshaw, R., Lane, R., Tanner, G., & Wofinden, D. (1998). *Levels of acitivity relating to safer routes to school type projects and green transport plans*. London: University of Westminster and Social and Transport Research Services.

British Standards Institution. (2008). *PAS 500: 2008—National specification for workplace travel plans*, London, UK.
Brockman, R., & Fox, K. R. (2011). Physical activity by stealth? The potential health benefits of a workplace transport plan. *Public Health, 125*, 210–216.
Brög, W., Erhard, E., Ker, I., Ryle, J., & Wall, R. (2009). Evaluation of voluntary travel behaviour change: Experiences from three continents. *Transport Policy, 16*, 281–292.
Cairns, S., Davis, A., Newson, C., & Swiderska, C. (2002). *Making travel plans work: Research report*. London, UK: Department for Transport.
Cairns, S., & Newson, C. (2006). *Making school travel plans work: Effects, benefits and success factors at English schools*, Association for European Transport and contributors.
Cairns, S., Newson, C., & Davis, A. (2010). Understanding successful workplace travel initiatives in the UK. *Transportation Research Part A: Policy and Practice, 44*(7), 473–494.
Cairns, S., Sloman, L., Newson, C., Anable, J., Kirkbride, A., & Goodwin, P. (2004). *Smarter choices—Changing the way we travel*. UK: Department for Transport.
Cairns, S., Sloman, L., Newson, C., Anable, J., Kirkbride, A., & Goodwin, P. (2008). Smarter choices: Assessing the potential to achieve traffic reduction using 'soft measures'. *Transport Reviews, 28*(5), 593–618.
Coleman, C. (2000). Green commuter plans and the small employer: An investigation into the attitudes and policy of the small employer towards staff travel and green commuter plans. *Transport Policy, 7*, 139–148.
Cooper, B., & Meiklejohn, D. (2003). *A new approach for travel behaviour change in universities*. Paper presented to 26th Australasian Transport Research Forum (ATRF), Wellington, New Zealand.
Currie, G., & Delbosc, A. (2011). Travel demand management for the summer olympic games. *Transportation Research Record: Journal of the Transportation Research Board, 2245*, 36–48.
Curtis, C., & Holling, C. (2004). Just how (travel) smart are Australian universities when it comes to implementing sustainable travel? *World Transport Policy and Practice, 10*(1), 22–33.
Davison, L., Enoch, M., & Ison, S. (2010). *European experience of travel plans: An expert perspective*. Paper presented to 12th World Conference in Transportation Research, Lisbon, Portugal.
DeGruyter, C., Rao, D., & Meiklejohn, D. (2005). *Tools for travel behaviour change*. Paper presented to 28th Australasian Transport Research Forum (ATRF), Sydney, Australia.
Department for Transport. (2002). *Using the planning process to secure travel plans: Best practice guidance for local authorities, developers and occupiers*. London, UK: Department for Transport.
Department for Transport. (2005). *Making residential travel plans work: Guidelines for new development*, London, UK.
Department for Transport. (2007). *Making residential travel plans work*, London, UK.
Department for Transport. (2009). *Good practice guidelines: Delivering travel plans through the planning process*. London, UK: Department for Transport.
Department of Infrastructure. (2008). *TravelSmart travel planning guide*, Victoria, Australia.
Department of Transport. (undated-a). *Travel planning case study: Love living local—City of Darebin*. Victoria, Australia: Department of Transport.
Department of Transport. (undated-b). *Travel planning case study: School travel planning on the Bellarine Peninsula—City of Greater Geelong*. Victoria, Australia: Department of Transport.
Department of Transport. (undated-c). *Travel planning case study: Transport Swinburne and corporate Boroondara—City of Boroondara*. Victoria, Australia: Department of Transport.
Department of Transport, Planning and Local Infrastructure. (2014a). *A guide to the planning system*. Retrieved September 25, 2014, from http://www.dpcd.vic.gov.au/planning/.
Department of Transport, Planning and Local Infrastructure. (2014b). *Planning schemes*. Retrieved September 25, 2014, from http://www.dpcd.vic.gov.au/planning/planningschemes.

Di Pietro, G., & Hughes, I. (2003). *TravelSMART schools—There really is a better way to go!* Paper presented to 26th Australasian Transport Research Forum (ATRF), Wellington, New Zealand.

Dill, J. (1998). Mandatory employer-based trip reduction: What happened? *Transportation Research Record: Journal of the Transportation Research Board, 1618*, 103–110.

Enoch, M. (2012). *Sustainable transport, mobility management and travel plans*. Surrey, England: Ashgate Publishing Limited.

Enoch, M., & Ison, S. (2008). *Expert perspectives on the past, present and future of travel plans in the UK: Research report*. Department for Transport and the National Business Travel Network.

Enoch, M., & Potter, S. (2003). Encouraging the commercial sector to help employees change their travel behaviour. *Transport Policy, 10*, 51–58.

Enoch, M., & Rye, T. (2006). Travel plans: Using good practice to inform future policy. In B. Jourquin, P. Rietveld, & K. Westin (Eds.), *Towards better performing transport networks* (pp. 157–177). London, UK: Routledge.

Ferguson, E. (2000). *Travel demand management and public policy*. England: Ashgate Publishing.

Fraser, J., & Addison, L. (2002). Travel plans—A greenwash or an effective planning mechanism. *Town and Country Planning*, 275–277.

Giuliano, G., Hwang, K., Perrine, D., & Wachs, M. (1991). *Preliminary evaluation of regulation XV of the south coast air quality management district*. Berkeley, California, US: University of California Transportation Center.

Guiver, J., & Stanford, D. (2014). Why destination visitor travel planning falls between the cracks. *Journal of Destination Marketing & Management*.

Halling, B., & Mayes, M. (2011). *'Workin' it—Making smarter travel at work smarter!* Paper presented to 34th Australasian Transport Research Forum (ATRF), Adelaide, Australia.

Hamre, A., & Buehler, R. (2014). Commuter mode choice and free car parking, public transportation benefits, showers/lockers, and bike parking at work: Evidence from the Washington, DC Region. *Journal of Public Transportation, 17*(2), 67–91.

Hancock, L., & Nuttman, S. (2013). Engaging higher education institutions in the challenge of sustainability: Sustainable transport as a catalyst for action. *Journal of Cleaner Production, 62*, 62–71.

Harrison, J. (2003). Travel plans and the planning system. *Journal of Planning and Environment Law*, 397–403.

Hendricks, S. (2005). Effectiveness of programs for work site trip reduction: The influence of organisational culture. *Transportation Research Record: Journal of the Transportation Research Board, 1924*, 207–214.

Hendricks, S. (2008). Four challenges to incorporating transportation demand management into the land development process. *Transportation Research Record: Journal of the Transportation Research Board, 2046*, 30–36.

Hendricks, S., & Georggi, N. (2007). Documented impact of transportation demand management programs through the case study method. *Journal of Public Transportation, 10*(4), 79–98.

Higgins, T. (1996). How do we know employer-based transportation demand management works? The need for experimental design. *Transportation Research Record: Journal of the Transportation Research Board, 1564*, 54–59.

Hill, J. (2013). *Proposed modifications to the regional transportation impact study guidelines—Public meeting, report P-13-031*. Ontario, Canada: Region of Waterloo.

Hinckson, E., & Badland, H. (2011). School travel plans: Preliminary evidence for changing school-related travel patterns in elementary school children. *American Journal of Health Promotion, 25*(6), 368–371.

Hinckson, E., Duncan, S., & Badland, H. (2009). Can school travel plans change the way children commute? *Journal of Science and Medicine in Sport,* 12.

Holzer, P. (2004). *Attitudes of private developers towards travel demand management*. Student Internship Report, Swinburne University.

Hooi, J. (2012). Travel smart: You don't have to go to work all together. *The Business Times*.

Howlett, R., & Watson, T. (2010). *Travel planning in Victoria—A new strategic approach to sustaining communities*. Paper presented to 33rd Australasian Transport Research Forum (ATRF), Canberra, Australia.

Ison, S., Merkert, R., & Mulley, C. (2014). Policy approaches to public transport at airports—Some diverging evidence from the UK and Australia. *Transport Policy, 35*, 265–274.

Ison, S., & Rye, T. (2008). *The implementation and effectiveness of transport demand management measures: An international perspective*. Hampshire, UK: Ashgate.

Jollon, M. (2013). *Best practices in integrating travel plans with the development approval process: Examples from the USA*. Paper presented to Planning Institute of Australia (PIA) 2013 National Congress, Canberra, Australia.

Khagram, A. (2013a). *Travel plan monitoring officer report: Covering autumn 2012*. UK: WestTrans.

Khagram, A. (2013b). *Travel plan monitoring officer report: Covering Spring 2013*. UK: WestTrans.

Khagram, A. (2014). The five cards that will ensure a travel plan is more than just a bluff. *TransportXtra*.

Khandokar, F., Ryley, T., Ison, S., & Price, A. (2013). *A survey on hospital travel plans in England*. Paper presented to 92nd Transportation Research Board (TRB) Annual Meeting, Washington DC.

Llewellyn, R., Tricker, R., & Paton, D. (2014). Travel plans: A critical comparison of the application of land use planning processes in England and Scotland. *Transport*, iFirst, 1–13.

Lopez-Aqueres, W. (1993). Conceptual framework to study the effectiveness of employer trip-reduction programs. *Transportation Research Record: Journal of the Transportation Research Board, 1404*, 55–63.

Mammen, G., et al. (2013). Active school travel: An evaluation of the Canadian school travel planning intervention. *Preventive Medicine*.

Mammen, G., Stone, M., Buliung, R., & Faulkner, G. (2014). School travel planning in Canda: Identifying child, family and school-level characteristics associated with travel mode shift from driving to active school travel. *Journal of Transport & Health*.

Marsden Jacob Associates. (2011). *Evaluation of the TravelSmart local government and workplace programs*. Perth, Australia: WA Department of Transport.

McFadden, H., DeGruyter, C., & Watson, T. (2006). *Widening the scope: Delivery of travel behaviour change initiatives by local government in Victoria*. Paper presented to Australian Institute of Traffic Management and Planning (AITPM) National Conference Melbourne, Australia.

Meiklejohn, D., & Wake, D. (2007). *A tale of two cities: Workplace travel plan programs in Melbourne and Perth*. Paper presented to 30th Australasian Transport Research Forum, Melbourne, Australia.

Mele, L. (2013). *TDM membership by Condo corporation*. 8 July 2013, Email communication to Pat Fisher (City of Waterloo, Ontario).

Melia, S. (2009). *Potential for carfree development in the UK*. PhD Thesis, University of West England.

Melia, S., Barton, H., & Parkhurst, G. (2013). Potential for carfree development in the UK. *Urban Design and Planning, 166*(DP2), 136–145.

Miller, E. (1995). Truth about employer trip reduction programs. *TDM Review*, 3(2), 14–16.

Moghtaderi, F., Burke, M., & Dodson, J. (2012). *A systematic review of children's travel behaviour change programs in Australia*. Paper presented to 35th Australasian Transport Research Forum (ATRF), Perth, Australia.

Morris, D., Enoch, M., Pitfield, D., & Ison, S. (2009). Car-free development through UK community travel plans. *Urban Design and Planning*, 162(DPI), 19–27.

Myers, K. (2005). *Travel behaviour change initiatives: A local government's innovations*. Paper presented to 28th Australasian Transport Research Forum (ATRF), Sydney, Australia.

Newson, C., Cairns, S., & Davis, A. (2010). *Making school travel plans work: Experience from English case studies*, Transport for Quality of Life.

NSW Government. (2011). *Active travel: Optus relocation.* Premier's council for active living. Retrieved September 25, 2014, from http://www.pcal.nsw.gov.au/case_studies/optus.

NZ Transport Agency. (2011). *Workplace travel plan guidelines*. Wellington, New Zealand: New Zealand Government.

Orski, C. K. (1993). *Evaluation of employee trip reduction programs based on California's experience with regulation XV.* Paper presented to 1993 ITE International Conference.

Pagano, A., & Verdin, J. (1997). Employee trip reduction without government mandates: Cost and effectiveness estimates from Chicago. *Transportation Research Record: Journal of the Transportation Research Board, 1598*, 43–48.

Petrunoff, N., Rissel, C., Wen, L. W., Xu, H., Meiklejohn, D., & Schembri, A. (2013). Developing a hospital travel plan: Process and baseline findings from a western Sydney hospital. *Australian Health Review, 37*(5), 579–584.

Potter, S., & Enoch, M. (2007). Mobility management in organisations. In J. P. Warren (Ed.), *Managing transport energy: Power for a sustainable future* (pp. 93–120). Oxford, UK: Oxford University Press.

RAC. (2014). *Travel planning for new developments: Advice for local governments.* Perth, Western Australia: RAC Mobility Bulletin #01. Retrieved September 25, 2014, from http://rac.com.au/.

Richter, J., Friman, M., & Gärling, T. (2011). Soft transport policy measures: Gaps in knowledge. *International Journal of Sustainable Transportation, 5*(4), 199–215.

Roby, H. (2010a). Can travel plans escape the planning Ghetto?' *Town and Country Planning*.

Roby, H. (2010b). *Using innovation and business models to analyse the organisational embedding of travel plans*. PhD Thesis, Open University.

Roby, H. (2010c). Workplace travel plans: Past, present and future. *Journal of Transport Geography, 18*, 23–30.

Rose, G., & Ampt, E. (2001). Travel blending: An Australian travel awareness initiative. *Transportation Research Part D: Transport and Environment, 6*(2), 95–110.

Rye, T. (1997). *The implementation of workplace transport demand management in large organisations*. PhD Thesis, Nottingham Trent University.

Rye, T. (1999a). Employer attitudes to employer transport plans: A comparison of UK and Dutch experience. *Transport Policy*, 6, 183–196.

Rye, T. (1999b). Employer transport plans—A case for regulation? *Transport Reviews, 19*(1), 13–31.

Rye, T., Green, C., Young, E., & Ison, S. (2011a). Using the land-use planning process to secure travel plans: An assessment of progress in England to date. *Journal of Transport Geography, 19*(2), 235–243.

Rye, T., Welsch, J., Plevnik, A., & de Tomassi, R. (2011b). First steps towards cross-national transfer in integrating mobility management and land use planning in the EU and Switzerland. *Transport Policy, 18*, 533–543.

Seggerman, K., & Hendricks, S. (2005). *Incorporating TDM into the land development process*. State of Florida: Department of Transportation.

Smith, L. (2010). School travel plans: How successful are they? *Traffic Engineering and Control*, 189–193.

Steer Davies Gleave. (2001). *Take-up and effectiveness of travel plans and travel awareness campaigns: Final report*. London, UK: Department of the Environment, Transport and the Regions.

Stewart, J. (1994). Reducing drive-alone rates at small employer sites: Costs and benefits of local trip reduction ordinances—Pasadena towers case study. *Transportation Research Record: Journal of the Transportation Research Board, 1433*, 159–163.

Sullivan, C., & Percy, A. (2008). *Evaluating changes associated with workplace and school travel plans—Something old, something borrowed, something new*. Paper presented to 31st Australasian Transport Research Forum (ATRF), Gold Coast, Queensland, Australia.

Thom, A. (2009). *Behaviour change towards sustainable travel in Perth—The TravelSmart workplace program*. Paper presented to PATREC Research Forum, Curtin University Perth, Australia.

Transport for London. (2008). *Guidance for residential travel planning in London*. London, UK: Transport for London.

Transport for London. (2011a). *ATTrBuTE v3 user guide*, London, UK. Retrieved September 25, 2014, from http://www.attrbute.org.uk/.

Transport for London. (2011b). *Travel planning for new development in London—Incorporating deliveries and servicing*. UK: Transport for London.

Travel Plan Services Ltd. (2013). *Newcastle great park framework travel plan implementation: Annual progress and monitoring report*, Wakefield, UK.

TRICS. (undated). *UK standard assessment method for travel plan impacts: Advice for developers and planning authorities*, UK. http://www.trics.org/.

Van Malderen, L., et al. (2013). Exploring the profession of mobility manager in Belgium and their impact on commuting. *Transportation Research Part A, 55*, 46–55.

Vanoutrive, T., Van Malderen, L., Jourquin, B., Thomas, I., Verhetsel, A., & Witlox, F. (2010). Mobility management measures by employers: Overview and exploratory analysis for Belgium. *European Journal of Transport Infrastructure and Research, 10*(2), 121–141.

Wake, D. (2012). *Engaging Perth workplaces for sustainable transport: An evaluation of the TravelSmart workplace program*. Paper presented to 35th Australasian Transport Research Forum (ATRF), Perth, Australia.

Wake, D., Thom, A., & Cummings, R. (2010). *Evaluating workplace travel plans*. Paper presented to 33rd Australasian Transport Research Forum (ATRF), Canberra, Australia.

Wiblin, S. (2010). *Integrating travel behaviour change for workers, shoppers and residents at an outer suburban centre*. Paper presented to 33rd Australasian Transport Research Forum (ATRF), Canberra, Australia.

Wiblin, S., Mulley, C., & Ison, S. (2012). *Precinct wide travel plans—Learnings from Rouse Hill Town Centre*. Paper presented to 35th Australasian Transport Research Forum (ATRF), Perth, Australia.

Winters, P., Perez, R., Joshi, A., & Perone, J. (2005). Work site trip reduction model and manual. *Transportation Research Record: Journal of the Transportation Research Board, 1924*, 197–206.

Woodruff, A., & Hui, C. (2010). *Integrating planning in activity centres: Influencing change across all travel purposes*. Paper presented to 33rd Australasian Transport Research Forum (ATRF), Canberra, Australia.

Wright, L. (2005). *Sustainable Transport: A Sourcebook for Policy-makers in Developing Cities Module 3e Car-Free Development, Deutsche Gesellschaft fur Technische Zusammenarbeit (GTZ) GmbH*. Eschborn, Germany: Federal Ministry for Economic Cooperation and Development.

WSP. (2014). *Does car ownership increase car use? A study of the use of car parking within residential developments in London*. UK: Commissioned by the Berkeley Group.

Wynne, L. (2013). *Travel planning and the land-use planning system: Understanding the effectiveness of travel planning requirements in NSW development control plans*. Unpublished Masters Thesis, University of New South Wales, Australia.

Yeates, S., & Enoch, M. (2012). *Travel plans from the developer perspective*. Paper presented to 91st Transportation Research Board (TRB) Annual Meeting, Washington DC.

Young, R., & Luo, R. (1995). Five-year results of employee commute options in southern California. *Transportation Research Record: Journal of the Transportation Research Board*, 1496, 191–198.

Chapter 3
Theoretical Foundations

3.1 Introduction

The previous chapter provided a literature review on travel plans and their application to new developments. In doing so, it identified a set of research gaps and opportunities that will become the focus of this thesis in subsequent chapters.

The aim of this chapter is to describe the theoretical foundations for this research by providing a literature review of both **implementation theory** and **planning enforcement theory** (Fig. 3.1). These theories strongly align with the research aim of identifying opportunities to enhance the implementation and subsequent effectiveness of travel plans for new residential developments.

Implementation theory provides guidance on the effective implementation of programs and policies, which can be directly applied to travel plans. By contrast, planning enforcement theory suggests suitable approaches for achieving planning compliance which can provide important lessons for travel planning in the context of new developments. The theories are particularly relevant to the research given the issues associated with implementing travel plans at new residential developments and the difficulties experienced by local government in enforcing them through the land use planning process.

This chapter begins with a review of the literature on implementation theory, focusing on both top-down and bottom-up approaches to implementation. It then describes planning enforcement theory, covering two main approaches: systematic enforcement and facilitative enforcement. The chapter concludes with a discussion of the theoretical implications for exploring the use of travel plans for new residential developments.

C. De Gruyter, *Travel Plans for New Residential Developments: Insights from Theory and Practice*, Springer Theses,
DOI 10.1007/978-981-10-2092-6_3

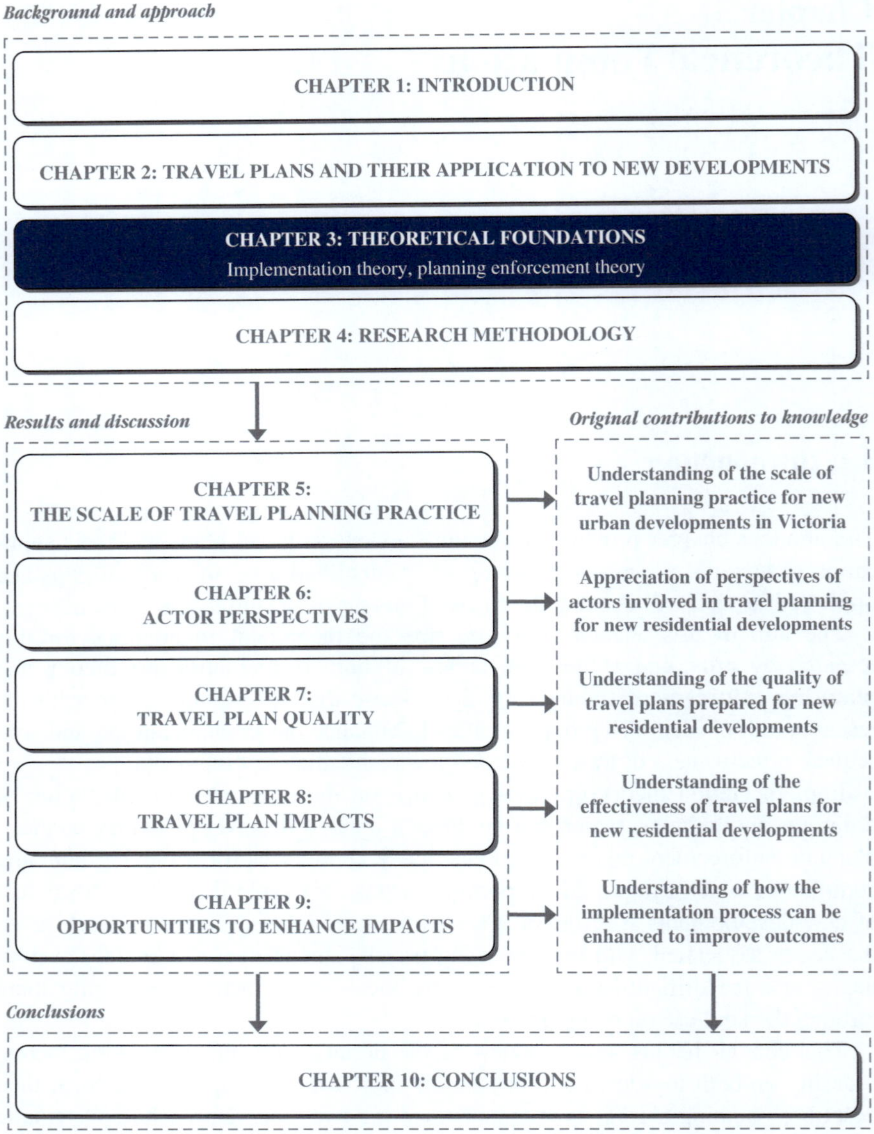

Fig. 3.1 Position of Chap. 3 in the thesis structure

3.2 Implementation Theory

Implementation theory developed out of the public policy field and provides valuable guidance on conditions for effective implementation and frameworks for conceptualising the policy implementation process (Mazmanian and Sabatier 1981). A distinction is often drawn between two fundamentally different approaches to implementation:

1. **Top-down approach**: a focus on program effectiveness and the ability to control the behaviour of implementers and target groups (Sabatier 1986)
2. **Bottom-up approach**: a focus on mapping the strategies of actors concerned with a particular policy issue or problem (Elmore 1979).

This section describes and compares these approaches to implementation.

3.2.1 Top-down Approach to Implementation

The top-down approach to implementation emphasises the ability of centrally-based decision makers to develop unequivocal policy objectives and control the implementation process (Pülzl and Treib 2007). Sabatier and Mazmanian (1980) identified a set of 17 variables that affect the implementation process from a top-down perspective. These variables are presented as a conceptual framework in Fig. 3.2, where they are grouped into tractability, statutory and non-statutory categories.

The 17 variables were also synthesised by Sabatier and Mazmanian (1981) into a shorter list of six sufficient and generally necessary conditions for effective implementation:

1. Policy objectives are clear and consistent
2. Program is based on an adequate causal theory (link between the problem and solution)

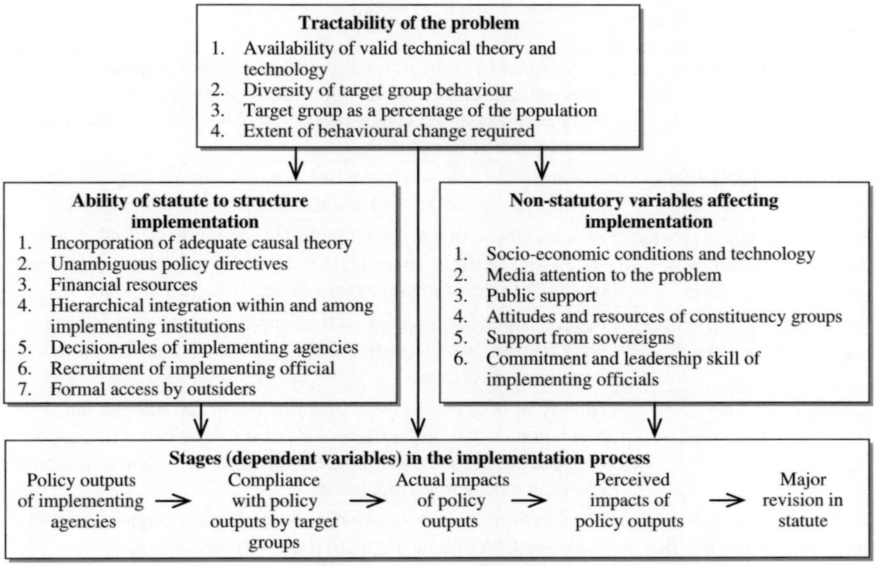

Fig. 3.2 Variables affecting the implementation process from a top-down perspective. *Source* Sabatier and Mazmanian (1980)

3. Implementation process is legally structured to enhance compliance
4. Implementing officials are both committed and skilful
5. Interest groups and sovereigns are supportive
6. Changes in socioeconomic conditions do not undermine political support or causal theory.

The first three conditions can be dealt with by the initial policy decision or regulation, whereas the latter three conditions are largely the result of external political and economic factors during the implementation process (Sabatier and Mazmanian 1981).

Also in line with the top-down approach, O'Toole (1986) set out a number of principles to guide successful implementation based on a review of top-down multi-actor implementation literature, while Gunn (1978) identified ten preconditions for perfect implementation.

Applications of top-down implementation theory to travel plans have been limited to date. Only two studies have applied the theory, both within the context of workplace travel plans. Firstly, the six top-down conditions for effective implementation developed by Sabatier and Mazmanian (1981) were applied by Marzotto et al. (2000) to examine workplace travel plans mandated by Regulation XV in the United States. A summary is presented in Table 3.1. This application helped to explain a number of implementation issues associated with Regulation XV, namely a lack of common understanding of the program's objectives and the election of a conservative, anti-regulation congress.

Table 3.1 Assessment of conditions for effective implementation (Sabatier and Mazmanian 1981) based on workplace travel plans mandated through Regulation XV in the United States

Conditions for effective implementation	Assessment based on workplace travel plans in the US
1. Clear and consistent objectives	Some states thought air quality was the primary objective and so assigned implementation to departments of the environment, others thought it was traffic congestion and so assigned implementation to departments of transport
2. Adequate causal theory	Not everyone believed in the link between commuting and air pollution. Regulation XV did not affect non-work related driving
3. Implementation process legally structured to enhance compliance	EPA had the authority to withhold federal highway funds from states who failed to promulgate acceptable regulations. However, program implementation never reached this critical point
4. Committed and skilful implementing officials	Federal officials disagreed with each other, while state agencies were frequently understaffed and faced opposition from state legislatures and governors
5. Support of interest groups and sovereigns	Quality of life coalition was instrumental in formulating the regulation as part of the Clean Air Act but provided little support to agency staff to fend off assaults from the economic development coalition during implementation
6. Changes in socio-economic conditions that do not undermine political support or causal theory	The election of a conservative, anti-regulation Congress in 1994 made the EPA and the regulation easy targets to attack

Source Author's adaptation based on Marzotto et al. (2000)

Secondly, the ten preconditions for perfect implementation developed by Gunn (1978) were applied by Ison and Rye (2003) to workplace travel plans in the United Kingdom (see Table 3.2). They note that while perfect implementation is virtually impossible in the real world, Gunn's preconditions do provide a

Table 3.2 Relevance of preconditions for perfect implementation (Gunn 1978) to workplace travel planning in the United Kingdom

Preconditions for perfect implementation	Relevance to workplace travel plans in the UK
1. Circumstances external to the implementing agency do not impose crippling constraints	External pressures (e.g. traffic congestion) can in fact motivate an organisation to develop a travel plan
2. Adequate time and significant resources are made available to the program	Little hope that a travel plan can achieve change if there is no one to implement it. Financial resources for travel plans can vary widely
3. Not only are there no constraints in terms of overall resources but also that, at each stage of the implementation process, the required combination of resources is actually available	Persuading senior management that resources for implementing a travel plan are required can be a challenging task
4. The policy to be implemented is based upon a valid theory of cause and effect	While clear evidence exists that a travel plan can reduce the number of car trips to a workplace, little evidence is available that *off-site* congestion can be reduced—this lack of evidence of cause and effect may remain a barrier to the adoption of travel plans
5. The relationship between cause and effect is direct and there are few, if any, intervening links	Cause and effect linkages within a travel plan are normally relatively simple—there is no great complex chain of causality
6. There is a single implementing agency which need not depend upon other agencies for success or, if other agencies must be involved, that the dependency relationships are minimal in number and importance	Travel plan coordinator may depend on other organisations, or other departments within their own organisation, for the implementation of some actions
7. There is complete understanding of, and agreement upon, the objectives to be achieved; and that these conditions persist throughout the implementation process	Organisations can generally specify objectives for their travel plan, although setting targets may be more difficult without knowing what travel plans have achieved at other locations
8. In moving towards agreed objectives it is possible to specify, in complete detail and perfect sequence, the tasks to be performed by each participant	Nature of the travel plan process not always clearly appreciated by organisations, partly because travel plans are a novel concept. This can lead to incorrect or non-specification of tasks
9. There is perfect communication among, and coordination of, the various elements or agencies involved in the program	While cross-departmental working groups are normally set up to implement and monitor travel plans in organisations, it may be difficult to fully coordinate activities as the travel plan is normally 'driven' by one or two departments
10. Those in authority can demand and obtain perfect obedience	Travel plans depend on voluntary changes in travel behaviour—it is impossible for an employer to require employees to commute in a particular way

Source Author's adaptation based on Ison and Rye (2003)

useful way to evaluate the implementation process. Of the ten preconditions, Ison and Rye (2003) note that external circumstances (precondition one), cause and effect theory (precondition four), having a single implementing agency (precondition six) and perfect communication (precondition nine) are the most important to successful workplace travel planning, based on their own experience with implementation.

While the top-down approach provides clear guidance on conditions for effective implementation, it has been argued that it incorrectly assumes a direct causal link between a policy and its outcomes, with little regard for the influence of implementers (Pülzl and Treib 2007). This gave rise to the bottom-up approach, as discussed in the next section.

3.2.2 Bottom-up Approach to Implementation

The bottom-up approach places a greater focus on implementers and recognises that policy is only one influence, and perhaps only a minor one, on the behaviour of implementers and subsequent target groups (Elmore 1979). A key premise supporting the bottom-up approach is that implementers, also termed 'street-level bureaucrats', have considerable discretion at their disposal as to how they interpret and implement a policy and it is therefore appropriate to account for these influences (Lipsky 1971). For example, street-level bureaucrats typically employ a number of coping mechanisms and simplifications when implementing policy and this is often done with inadequate information and too little time to weigh up the merits of each option (Lipsky 1971). Furthermore, Sabatier (1986, p. 22) states that implementers can 'often deflect centrally-mandated programs towards their own ends'.

No studies have explicitly applied the bottom-up approach to travel plans. However, Marzotto et al. (2000) do note the following in the context of workplace travel plans mandated by Regulation XV in the United States:

> Failure to take into account the private [bottom-up] side of implementation and the linkages between the public and private sectors will result in poorly enforced and ineffective implementation, few policy outputs, little policy impact, and, ultimately, unsolved public problems (Marzotto et al. 2000, p. 119).

Furthermore, in the context of voluntary workplace travel plans in Australia, Askew (2011) notes that 'ultimately, it is the actors that propel and shape the process, and their roles and activities within the process should therefore be systematically represented.'

These findings are consistent with Elmore (1978, p. 209) who maintains that a 'frequent explanation of implementation failures is that those who implement programs are seldom included in decisions that determine the content of those programs'. The findings are also consistent with the literature review presented in

Chap. 2 of this thesis which identified ownership and engagement as key to the success of travel plans.

Unlike the top-down approach, the bottom-up approach is not based on a set of preconditions for successful implementation. Rather, practitioners generally focus on the multitude of actors who interact at the local level on a particular issue and the strategies used by these actors in pursuit of their objectives (Sabatier 1986). According to O'Toole (2007, p. 147), the number of actors involved in delivering a given policy can affect the probability of implementation success and that with 'sequential arrangements, adding more organizational units in a chain increases the number of possible roadblocks to action.'

However, the bottom-up is also not without its limitations. These include the tendency to focus only on the goals and strategies of local actors, and failing to take into account top-down related influences (Sabatier 1986). It is therefore useful to compare approaches to implementation, as discussed in the next section.

3.2.3 Comparing Approaches to Implementation

Table 3.3 provides a comparison of key characteristics of the top-down and bottom-up approaches to implementation. The top-down approach tends to take the perspective of central policy makers, with formal regulation used to prescribe a set of program outputs. This contrasts with the bottom-up approach which takes the perspective of decentralised street-level bureaucrats (or local implementers) who use both formal and informal methods to solve issues to local problems.

A key strength of the top-down approach is its ability to provide a clear set of conditions for effective implementation. These conditions are particularly useful when there is a focus on assessing the effectiveness of a program (Sabatier 1986). However, application of the top-down approach is limited in situations where there is no dominant agency, but rather a multitude of actors. Furthermore, it is 'likely to ignore, or at least underestimate, the strategies used by street level bureaucrats

Table 3.3 Comparison of top-down and bottom-up approaches to implementation

Characteristic	Top-down approach	Bottom-up approach
Policy decision-maker	Policymakers	Street-level bureaucrats
Starting point	Statutory language	Social problems
Structure	Formal	Both formal and informal
Process	Purely administrative	Networking, including administrative
Authority	Centralisation	Decentralisation
Output/outcomes	Prescriptive	Descriptive
Discretion	Top-level bureaucrats	Bottom-level bureaucrats

Source Paudel (2009)

and target groups to get around (central) policy and/or to divert it to their own purposes (Sabatier 1986, p. 30).

Conversely, the bottom-up approach is able to deal with policy areas involving a multitude of actors (O'Toole 1986). Also, given the focus is not on the attainment of formal policy objectives, all types of (unintended) consequences of a program can be analysed. However, the bottom-up approach can underestimate the ability of top-down variables to shape the institutional structure in which individuals operate, thereby indirectly influencing the goals and strategies of local actors (Sabatier 1986).

In discussing the relative merits and limitations of top-down and bottom-up approaches to implementation, Pülzl and Treib (2007) note that there is now general agreement among scholars that implementation is located on a continuum between central authority (top-down) and local autonomy (bottom-up). They suggest that the 'preferences of street-level bureaucrats and the negotiations within implementation networks have to be taken into account to the same extent as centrally defined policy objectives and efforts at hierarchical control' (Pülzl and Treib 2007, p. 100). In the context of this thesis, the top-down approach is relevant given that the effectiveness of travel plans for new residential developments is being explored. However, the bottom-up approach is also applicable given the multitude of actors involved in the travel planning process. It is therefore considered appropriate to utilise both top-down and bottom-up perspectives for studying the implementation of travel plans for new residential developments.

Efforts to combine the top-down and bottom-up approach have led to the development of the Advocacy Coalition Framework (ACF), originally proposed by Sabatier (1986). This states that policy subsystems, rather than a specific government organisation, are more useful for understanding public policy as they include all of the actors involved in the process (e.g. all levels of government, private businesses, and individual people), not just the street-level bureaucrats or government alone. This bottom-up focus is then combined with some of the more top-down influences, such as socio-economic conditions and legal instruments, to paint a more balanced picture of the policy implementation process (Sabatier 1986). The ACF assumes that 'actors can be aggregated into a number (usually one to four) of 'advocacy coalitions', each composed of actors from various governmental and private organizations who both (a) share a set of normative and causal beliefs and (b) engage in a non-trivial degree of co-ordinated activity over time' (Sabatier 1998, p. 103).

While the ACF provides a synthesis of the top-down and bottom-up approach, it is primarily focused on policy change (generally associated with a change in beliefs within coalitions), rather than implementation (Winter 2007). For this reason, it has limited relevance for understanding the implementation of travel plans for new residential developments so it is not considered further in this thesis.

This section has provided a literature review of implementation theory. The next section considers planning enforcement theory which also forms part of the theoretical foundations for the research.

3.3 Planning Enforcement Theory

Planning enforcement theory has received little attention in the research literature (Harris 2011). However, the theory does suggest two main approaches for achieving planning compliance:

1. **The systematic approach**: the uniform and strict application of rules, with legislative mechanisms used to deter violations (Prior 2000)
2. **The facilitative approach**: the use of incentives, negotiation, education and fostering of good working relationships to achieve compliance (Burby et al. 1998).

This section describes and compares these two approaches to planning enforcement. Despite the issues with enforcing travel plans through the planning system (as previously discussed in Chap. 2), planning enforcement theory is yet to be applied in the context of travel plans.

3.3.1 Systematic Approach to Planning Enforcement

The systematic approach to planning enforcement favours the use of legislation to deter violations such as through sanctions and fines. The ability to use systematic enforcement is considered to help in protecting the integrity of the planning system, particularly in cases of repeat and flagrant offenders (Harris 2010; Prior 2000).

Prior (2000) characterises systematic enforcement as a model which assumes that:

- Breaches of regulations are essentially intended
- Most perpetrators are aware of required rules and standards
- The threat of punitive sanctions is an essential deterrent to potential violation
- A comprehensive approach to enforcement is essential
- Rules are clear and unambiguous
- Regulators are effectively resourced and empowered
- Enforcement actions derive from reactions to violations.

However, such assumptions often turn out to be unwarranted (Burby et al. 1998; Prior 2000). This has given rise to the facilitative approach, as discussed in the next section.

3.3.2 Facilitative Approach to Planning Enforcement

The facilitative approach is centred on securing compliance, with punitive measures retained as a last resort. This approach favours the use of incentives, negotiation and education to assist offenders to comply with regulations (Burby et al.

1998; McKay 2003). The facilitative approach is based on the assumption that most breaches of regulations occur through ignorance and are therefore unintended (McKay 2003; Prior 2000). Harris (2011) argues that the facilitative approach can be well suited to situations where resources are limited (as they often are in planning enforcement), given that this approach is less resource intensive than systematic enforcement regimes.

Through a national survey of 819 local governments in the United States, Burby et al. (1998) found that enforcement is more likely to be effective with a facilitative approach. Their study identified four key ingredients to achieving successful compliance:

- An adequate number of technically competent staff
- Strong proactive leadership
- Adequate legal support
- A consistently strong effort to check building and development plans, inspect building and development sites, and provide technical assistance.

In the state of Victoria, Australia, local guidance on planning enforcement suggests that an emphasis should be placed on obtaining compliance rather than on prosecuting offenders (Planning Enforcement Officers Association Inc. 2007). In line with this approach, it is also recommended that training of enforcement officers should build skills in verbal and written communication, negotiation and conflict resolution (Victorian Auditor-General 2008).

3.3.3 Comparing the Systematic and Facilitative Approaches to Planning Enforcement

A conceptual representation of the approaches to planning enforcement is presented in Fig. 3.3. The facilitative approach mostly relates to the bottom layer of the pyramid where information, advice and negotiation are key features. Conversely, the systematic approach to enforcement is typically characterised by the top of the pyramid where legal action may be taken to prosecute offenders. In line with the theory and guidance on planning enforcement, most efforts should be directed towards the bottom of the pyramid, with 'prosecution' only used as a last resort when all other options are exhausted (Burby et al. 1998; McKay 2003; Planning Enforcement Officers Association Inc. 2007).

While the theory on planning enforcement supports the need for a facilitative approach, there is a clear role for systematic enforcement to deal with instances of repeat and flagrant offenders, and to protect the integrity of the planning system. As noted by Lai et al. (2007, p. 540):

> A compliance system becomes more effective when enforcement strategies are combined with appropriate management tactics. Whereas managerial problem solving is effective in handling most violations, more problematic failures have to be solved through

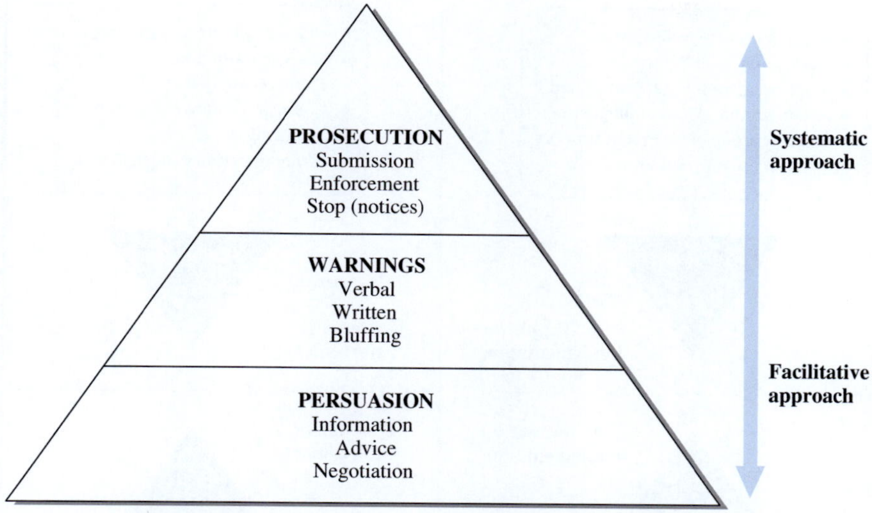

Fig. 3.3 Planning enforcement pyramid. *Source* Author's adaptation based on McKay (2003)

> enforcement with measures such as penalties. Such procedures for dealing with non-compliance are most effective when 'management' and 'enforcement' tactics are combined.

Furthermore, McKay (2003, p. 424) notes the benefits of combining the systematic and facilitative approaches to achieve compliance:

> …the complexity of the enforcement equation is such that it cannot be solved solely by legislative mechanisms. Key components in its resolution include adequate numbers of technically competent staff, strong proactive leadership, rigorous monitoring of planning conditions and the deployment of a facilitative enforcement strategy that fosters developers' commitment to comply with regulations.

Figure 3.4 provides a conceptual representation of implementation theory and planning enforcement theory. Key characteristics of the top-down and bottom-up approaches to implementation are shown, along with key features of the systematic and facilitative approaches to enforcement. The dashed lines dividing the approaches denote their consideration as distinct and separate entities.

Figure 3.4 also implies a commonality between top-down implementation and systematic enforcement, using inverted pyramids. With both of these approaches, an implicit assumption is made that a regulation, once enacted, is largely followed and successfully implemented (Prior 2000; Sabatier 1986). Similarly, the complementarity between bottom-up implementation and facilitative enforcement is also reflected in Fig. 3.4. The influence of street-level bureaucrats, typically associated with bottom-up implementation, has been noted by Prior (2000, p. 65) in the context of planning enforcement, maintaining that 'it is not the legislation per se that determines deviant behaviour, but rather the perspectives and priorities of the regulators… and the knowledge and motivations of the regulated.'

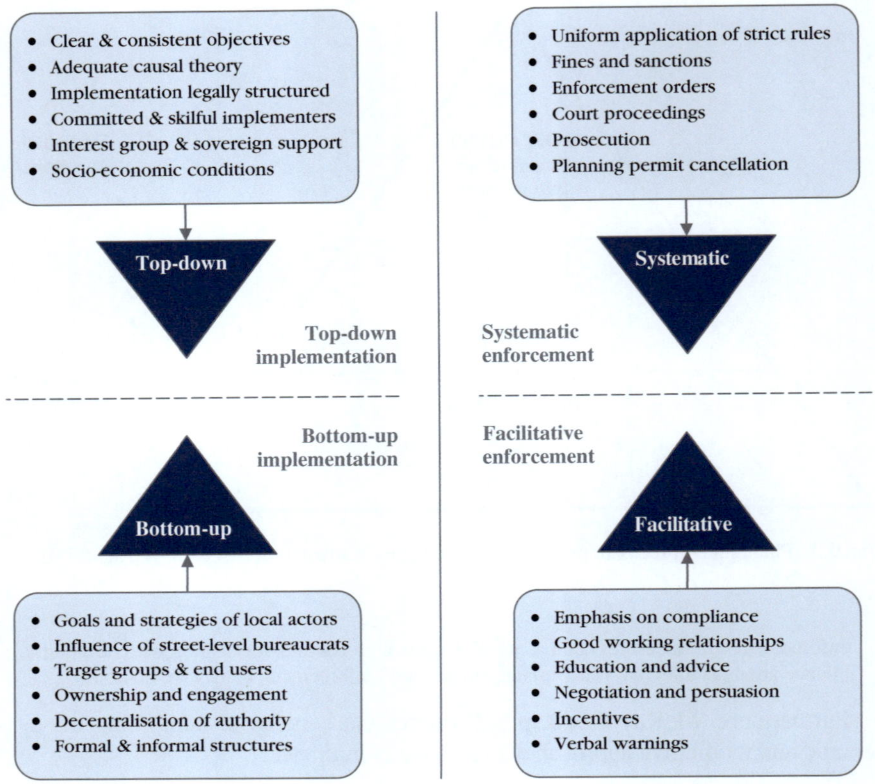

Fig. 3.4 Conceptualisation of implementation theory and planning enforcement theory. *Source* Author's synthesis

3.4 Conclusion

The aim of this chapter is to describe the theoretical foundations for this study by providing a literature review of both implementation theory and planning enforcement theory. An overview of the top-down and bottom-up approaches to implementation was presented, along with coverage of the systematic and facilitative approaches to planning enforcement.

Applications of the theories to travel plans have been limited to date, with only top-down perspectives taken to the implementation of workplace travel plans (Ison and Rye 2003; Marzotto et al. 2000). Planning enforcement theory is yet to be applied in the context of travel plans. The application of these theories to travel plans for new residential developments is needed to assist in identifying opportunities for enhancing their effectiveness, both in terms of approaches to their implementation and enforcement. This need is addressed in Chap. 9 of this thesis.

Furthermore, given the similarities between implementation theory and planning enforcement theory, an opportunity exists to advance towards an integrated theory of implementation and enforcement. This is discussed further in Chap. 9 where the findings from this research study are assessed in terms of the extent to which they support such an integrated theory.

The next chapter of this thesis outlines the research methodology for addressing the research gaps identified earlier in Chap. 2, including application of the theories described in this chapter.

References

Askew, M. (2011). *When travel matters: Opportunities, barriers and policy options for workplace travel management in Sydney*. Ph.D. thesis, Macquarie University, Sydney.

Burby, R., May, P., & Paterson, R. (1998). Improving compliance with regulations: Choices and outcomes for local government. *Journal of the American Planning Association, 64*(3), 324–334.

Elmore, R. (1978). Organizational models of social program implementation. *Public Policy, 26*(2), 185–228.

Elmore, R. (1979). Backward mapping: Implementation research and policy decisions. *Political Science Quarterly, 94*(4), 601–616.

Gunn, L. A. (1978). Why is implementation so difficult? *Management Services in Government, 33*, 169–176.

Harris, N. (2010). Discretion and expediency in the enforcement of planning controls. *Town Planning Review, 81*(6), 675–700.

Harris, N. (2011). Discipline, surveillance, control: A foucaultian perspective on the enforcement of planning regulations. *Planning Theory & Practice, 12*(1), 57–76.

Ison, S., & Rye, T. (2003). Lessons from travel planning and road user charging for policy-making: through imperfection to implementation. *Transport Policy, 10*, 223–233.

Lai, L., Yung, P., Li, R., & Ho, D. (2007). The private supply of and public demand for planning: Compliance with planning conditions in the absence of direct statutory enforcement measures. *Planning Practice and Research, 22*(4), 535–557.

Lipsky, M. (1971). Street-level bureaucracy and the analysis of urban reform. *Urban Affairs Quarterly, 6*, 391–409.

Marzotto, T., Burnor, V., & Bonham, G. (2000). *The evolution of public policy: Cars and the environment*. Colorado, US: Lynne Rienner Publishers Inc.

Mazmanian, D., & Sabatier, P. (1981). *Effective policy implementation.* Lexington Books, D.C. Heath and Company, US.

McKay, S. (2003). Sheriffs and outlaws: In pursuit of effective enforcement. *Town Planning Review, 74*(4), 423–443.

O'Toole, L. (1986). Policy recommendations for multi-actor implementation: An assessment of the field. *Journal of Public Policy, 6*(2), 181–210.

O'Toole, L. (2007). Interorganizational relations in implementation. In B. Peters & J. Pierre (Eds.), *Handbook of public administration* (pp. 142–152). London, UK: SAGE Publications Ltd.

Paudel, N. R. (2009). A critical account of policy implementation theories: Status and reconsideration. *Nepalese Journal of Public Policy and Governance, 25*(2), 36–54.

Planning Enforcement Officers Association Inc. (2007). *A guide to planning enforcement in Victoria*, PEOA, Victoria, Australia, viewed 25 September 2014. http://planning-enforcement.com/.

Prior, A. (2000). Problems in the theory and practice of planning enforcement. *Planning Theory & Practice, 1*(1), 53–69.

Pülzl, H., & Treib, O. (2007). Implementing public policy. In F. Fischer, G. Miller & M. Sidney (Eds.), *Handbook of public policy analysis: Theory, politics and methods.* US: Taylor & Francis Group.

Sabatier, P. (1986). Top-down and bottom-up approaches to implementation research: A critical analysis and suggested synthesis. *Journal of Public Policy, 6*(1), 21–48.

Sabatier, P. (1998). The advocacy coalition framework: Revisions and relevance for Europe. *Journal of European Public Policy, 5*(1), 98–130.

Sabatier, P., & Mazmanian, D. (1980). The implementation of public policy: A framework of analysis. *Policy Studies Journal, 8*(4), 538–560.

Sabatier, P., & Mazmanian, D. (1981) The implementation of public policy: A framework of analysis. In: D, Mazmanian, & P. Sabatier (Eds.), *Effective policy implementation.* US: D.C. Heath and Company.

Victorian Auditor-General. (2008). *Enforcement of planning permits.* Australia: Victoria.

Winter, S. (2007). Implementation perspectives: Status and reconsideration. In B. Peters & J. Pierre (Eds.), *Handbook of public administration* (pp. 131–141). London, UK: SAGE Publications Ltd.

Chapter 4
Research Methodology

4.1 Introduction

Chapter 2 provided a literature review of travel plans and their application to new developments. This identified a set of research gaps and opportunities as the focus for subsequent chapters. Chapter 3 then described the theoretical foundations for this research, namely implementation theory and planning enforcement theory. The alignment of these theories to the research aim was discussed, particularly in how they can assist in identifying opportunities to enhance the effectiveness of travel plans for new residential developments.

The aim of this chapter is to describe the approach taken to address the research gaps and opportunities, including the application and integration of implementation theory and planning enforcement theory (Fig. 4.1).

This chapter begins with a review of research methods followed by a description of the overall research approach. Limitations of the research approach are then discussed. The final section provides a set of concluding remarks and sets the scene for subsequent chapters of the thesis.

4.2 Review of Research Methods

Table 4.1 provides a summary of common research methods in terms of their objective, instrument, advantages and limitations. The types of research methods covered is not intended to be exhaustive, but rather to help provide context for the research approach that was adopted.

Interviews, focus groups and case studies are generally qualitative in nature, while surveys and secondary data analyses are more quantitative techniques.

C. De Gruyter, *Travel Plans for New Residential Developments: Insights from Theory and Practice*, Springer Theses,
DOI 10.1007/978-981-10-2092-6_4

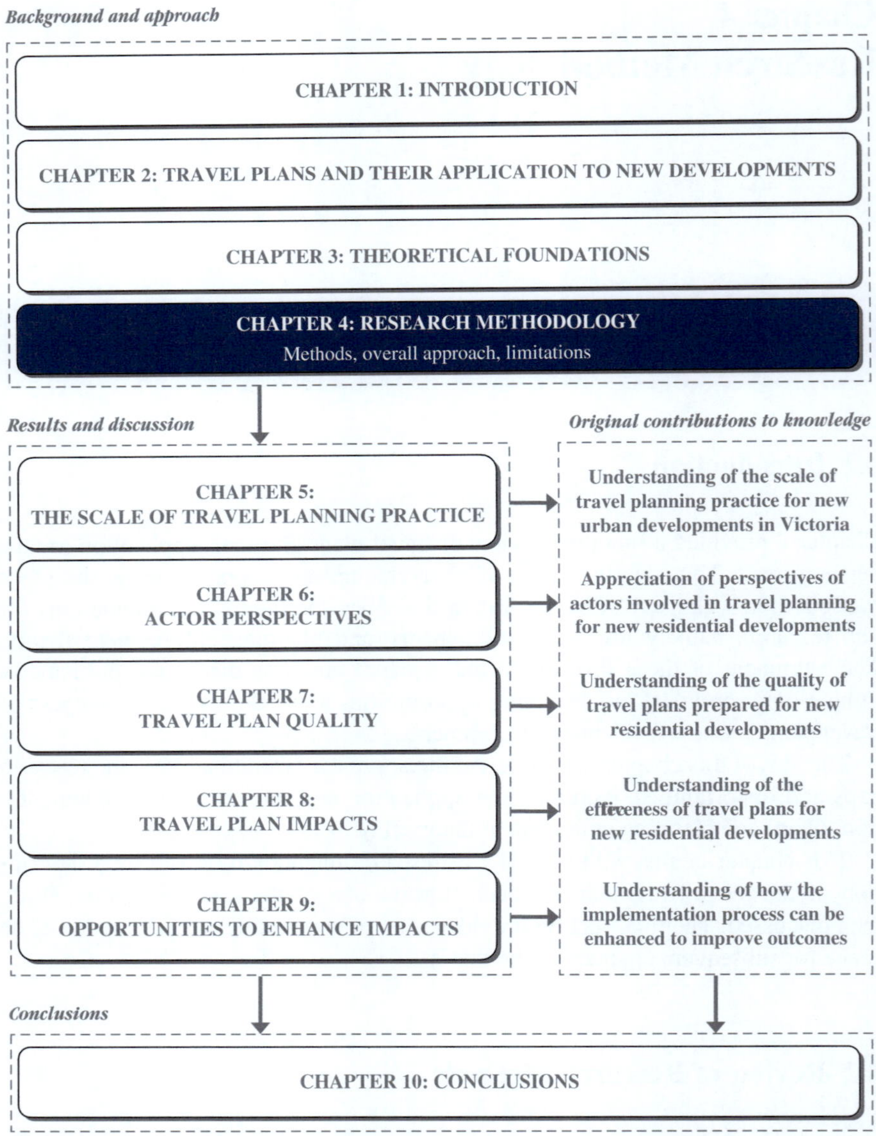

Fig. 4.1 Position of Chap. 4 in the thesis structure

Observation, document review and mixed-methods approaches can either take a qualitative or quantitative form depending on the specific objectives of the research (Bryman 2001).

As described in Table 4.1, each research method has a number of advantages and limitations. The extent to which these are present can vary depending on the specific context in which the research is being undertaken.

Table 4.1 Summary of research methods

Research method	Objective	Research instrument	Advantages	Limitations
Focus groups (qualitative)	To understand a range of opinions on a specific issue or to seek community norms	Discussion guide	• Group dynamics can offer additional insight over individual data collection • Can probe for additional information • Useful for exploring new topics and issues in depth • High emancipatory effect	• Potential for facilitator bias • Discussion may be dominated by a small number of participants • Time consuming to conduct, analyse and interpret findings • Cannot usually discuss sensitive topics
Interviews (qualitative)	To understand individual perceptions, beliefs, feelings and experiences	Interview guide	• Useful for asking 'why' and 'how' questions, and for understanding perceptions, beliefs and opinions • Useful for exploring issues in depth • Can probe for additional information • Non-verbal participant behaviour can offer additional insight	• Potential for interviewer bias • Potential for reflexivity bias (participant gives answer they think interviewer wants to hear) • Time consuming to conduct, analyse and interpret findings • Multiple interviews needed to identify a range of issues • May be intrusive for participants
Case studies[a] (qualitative)	To study a contemporary phenomenon within a real-life context	Case study protocol	• Can provides in-depth description of participants' experience • Helps to provide simple explanation or example of a situation	• Time consuming to collect, review and analyse data • Limited in breadth • Typically not generalisable to a population

(continued)

Table 4.1 (continued)

Research method	Objective	Research instrument	Advantages	Limitations
Observation (qualitative/quantitative)	To observe how others act and interact in natural settings	Observation guide	• Does not rely on willingness of participants to provide information • Can directly observe what participants do rather than what they say they do • Behaviour can be observed in its natural setting • Generally unobtrusive	• Potential for observer bias • Participants may act differently if they know they are being observed • Does not aid understanding of why participants behave the way they do • Privacy or access issues may exist
Mixed-methods[a] (qualitative/quantitative)	To explore a research problem using multiple research methods	Various	• Provides flexibility in addressing the research questions • Enables triangulation of findings and therefore greater confidence in results • Overcomes limitations of a single method	• Conclusions reached from different components may be incompatible • Can be relatively expensive and time consuming to conduct
Document review (qualitative/quantitative)	To identify themes, patterns and commonalities from existing documents	Coding frame	• Relatively inexpensive • Unobtrusive method • Can provide information that may not be directly observable • Longitudinal analyses often possible	• Documents may be out of date, incomplete or unavailable • Documents may be inaccurate or biased towards selected information • Can be time consuming to collect, review and analyse

(continued)

Table 4.1 (continued)

Research method	Objective	Research instrument	Advantages	Limitations
Survey (quantitative)	To elicit answers to 'how' and 'what' questions from respondents on a particular topic or set of topics	Questionnaire	• Relatively inexpensive • Can target respondents across a wide geographical area • Convenient for respondents in that they can complete a questionnaire when they want • Analysis is relatively quick • Useful for asking respondents 'how much', 'how often' and 'what proportion' type questions • Usually generalisable to a population	• Potential for low response rates • Some questions may be interpreted differently by others • Unable to probe for additional information • Difficult to know if the right person completed the questionnaire • Self-reporting may not reflect actual behaviour (social desirability bias) • Requires careful attention to questionnaire design
Secondary data analysis (quantitative)	To supplement existing findings or to explore a topic from a different angle	Data analysis technique/ program	• Relatively inexpensive and quick • Published data sources are usually of high quality • Longitudinal analyses often possible • Analyses across different geographies are often possible	• Lack of familiarity with data set • No control over data quality • May not contain all variables of interest • Data may be difficult to access

Source Author's synthesis of the literature based on Bryman (2001), Greenfield (2002), Hall (2008), Hennink et al. (2011), Mack et al. (2005) and Yin (2009)
[a]Considered more as a research strategy/approach than a research method, but included in the table for completeness

4.3 Overall Research Approach

The research gaps and opportunities identified in Chap. 2 are restated in Table 4.2. Their alignment with each research objective identified in Chap. 1 is also shown. In order to achieve the research objectives, five key research components (or tasks) were identified, as shown in Table 4.2. Key findings from each research component are reported throughout Chaps. 5–9.

Key inputs and outcomes of each research component are shown in Fig. 4.2. A mixed methods approach was adopted comprising surveys, interviews, document reviews and case studies. This approach was considered appropriate as a single method would not be capable of achieving all of the research objectives given their diverse nature. In accordance with Table 4.1, the mixed methods approach overcomes the limitation of using a single method, yet also enables triangulation of the findings and therefore greater confidence in the results (Bryman 2001).

A brief description of each research component is provided in the following sections.

4.3.1 Research Component 1: Online Survey of Victorian Councils

In addressing research objective 1, this research component involved an online survey of 36 (out of 79) Victorian councils to examine the scale and associated characteristics of travel planning practice for new urban developments in Victoria, Australia.

A survey was considered to be the most appropriate method for achieving the research objective given that answers to 'how much', 'how often' and 'what proportion' type questions were desired (Hennink et al. 2011). These types of questions are less suited to focus groups or interviews where topics are often more opinion based and discussed in greater depth (Mack et al. 2005). In accordance with Table 4.1, the survey method allowed respondents to be targeted across a wide geographical area, as is the case with the state of Victoria. It was also favourable for respondents as they could complete the survey at a time and place convenient to them. The survey could also be hosted online to facilitate ease of survey administration and data analysis.

Survey questions covered the extent to which travel plans had been required, reasons for requiring (and not requiring) them, planning mechanisms used, levels of monitoring undertaken, familiarity and experience with travel plans, perceptions of effectiveness, and the likelihood of requiring travel plans in the future.

Further detail on the aim, method and results associated with this research component is provided in Chap. 5.

Table 4.2 Linkages between research gaps, research opportunities, research objectives, research components and thesis chapters

Research gap →	Research opportunity →	Research objective[a] →	Research component →	Thesis chapter
There is no understanding of the scale and associated characteristics of travel planning practice for new developments in Australia *Identified in Sect.* 2.4	Examine the scale and associated characteristics of travel planning practice for new urban developments, using a case study from the state of Victoria	1. To examine the scale of practice in Victoria, Australia	1. Online survey of Victorian councils	Chapter 5: The scale of travel planning practice
No research has specifically explored the perspectives of the different actors involved in travel planning for new residential developments *Identified in Sect.* 2.3	Develop an appreciation for the perspectives of actors involved in travel planning for new residential developments, particularly aspects relating to implementation	2. To gain an appreciation for the perspectives of industry actors involved in their application	2. Interviews with industry representatives	Chapter 6: Actor perspectives
No formal assessment of the quality of travel plans prepared for new residential developments has been undertaken *Identified in Sect.* 2.5	Undertake a quantitative assessment of the quality of travel plans prepared for new residential developments to help identify their relative merits and potential areas for improvement	3. To evaluate their quality and effectiveness	3. Desktop assessment of travel plans	Chapter 7: Travel plan quality
Little research has been undertaken to appropriately quantify the effectiveness of travel plans in reducing car use at new residential developments, with no studies accounting for self-selection effects *Identified in Sect.* 2.6	Using a case study approach, evaluate the impacts of travel plans for new residential developments including self-selection effects to understand their effectiveness in reducing car use		4. Case studies of new residential developments	Chapter 8: Travel plan impacts

(continued)

Table 4.2 (continued)

Research gap →	Research opportunity →	Research objective[a] →	Research component →	Thesis chapter
No research has been undertaken to sufficiently explore implementation in the context of travel plans for new residential developments *Identified in Sect.* 2.5	Explore the implementation process associated with travel plans for new residential developments to identify opportunities to enhance effectiveness	4. To identify and assess opportunities for enhancing their implementation	5. Application and integration of implementation and planning enforcement theories	Chapter 9: Opportunities to enhance impacts

[a]Research objectives were identified in Chap. 1 (Sect. 1.3) and are framed in the context of travel plans for new residential developments

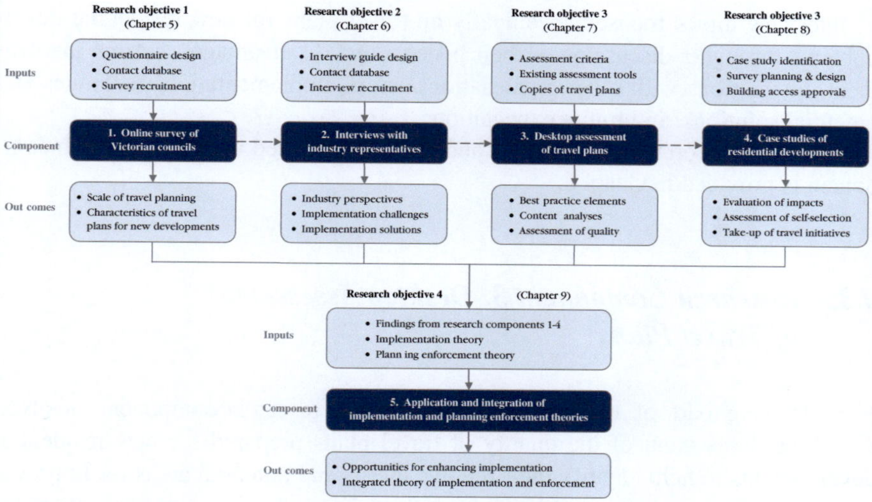

Fig. 4.2 Research approach in terms of inputs, components and outcomes

4.3.2 Research Component 2: Interviews with Industry Representatives

In addressing research objective 2, this research component involved a set of semi-structured interviews with 30 industry representatives to develop an appreciation for the perspectives of actors involved in travel planning for new residential developments. Industry representatives that were interviewed were based across 20 different organisations in Australia and the United Kingdom. The types of organisations included state and local government agencies, property development and management companies, and consultancies.

Interviews were considered to be the most appropriate method for achieving the research objective as perceptions, beliefs and experiences were being sought from industry representatives (Hennink et al. 2011). This is supported by Table 4.1 and is in contrast to a survey where answers to ‘how much’, ‘how often’ and ‘what proportion’ type questions are typically asked (Hennink et al. 2011). Focus groups were also not adopted for this research component due to the need to seek honest opinions that would not be biased by other participants. For example, a property developer or consultant may not speak with full candour if participating in a focus group with a government representative who is responsible for approving their planning application. A semi-structured approach to the interviews was chosen to retain flexibility, thereby allowing greater emphasis to be placed on some of the interview topics as required. Interview data was analysed using a set of codes that were developed inductively based on key themes that arose from the interview responses.

Interview topics focused specifically on travel plans for new residential developments through a discussion of their benefits and disadvantages, current involvement, interactions with other organisations/actors, implementation challenges and potential solutions, and future expectations.

Further detail on the aim, method and results associated with this research component is provided in Chap. 6.

4.3.3 Research Component 3: Desktop Assessment of Travel Plans

In addressing part of research objective 3, this research component involved a desktop assessment of the quality of travel plans prepared for new residential developments to help identify their relative merits and potential areas for improvement. This included the development and application of a quantitative assessment framework to 29 travel plans prepared for new residential developments in Victoria. It also included a content analysis to identify key characteristics of the travel plans.

In reference to Table 4.1, the method adopted for this research component was a set of document reviews. This was considered the most appropriate method as process-based aspects of travel plans are not always directly observable, thereby limited the potential for observational techniques to be used (Bryman 2001). Furthermore, patterns and commonalities in the travel plans were being assessed, which cannot easily be undertaken using other research methods. While a survey or set of interviews could provide some indication of the quality of travel plans prepared for new residential developments, these methods would not allow for a detailed analysis of travel plan content to be made.

Further detail on the aim, context, method and results associated with this research component is provided in Chap. 7.

4.3.4 Research Component 4: Case Studies of New Residential Developments

In addressing the remainder of objective 3, this research component involved a case-control design based on four case studies of new residential developments with travel plans, to understand their effectiveness in reducing car use. Matching control sites, similar in nature to the case study sites but without travel plans, were used as a comparison for evaluating the impacts of the travel plans. The case study and control sites are all located in Melbourne.

This research component adopted a mixed-methods approach using the following methods:

- **Observation**: multi-modal person trip counts at each case study and control site to provide information on transport mode shares and vehicle trip generation rates, plus car and bicycle parking surveys at each site to provide information on parking demand, supply and utilisation.
- **Survey**: a travel survey of residents at each site with a focus on travel characteristics, attitudes and preferences towards different forms of transport, and demographics. This was combined with a technique known as Propensity Score Matching (PSM) to assess the extent of self-selection among residents living at the case study sites.
- **Secondary data analysis**: use of published vehicle trip generation rates and regional travel survey data to provide a supplementary comparison to the multi-modal counts and parking surveys conducted at the case study and control sites.

Each of these research methods individually contribute towards developing an understanding of the effectiveness of travel plans in reducing car use at new residential developments. The case study approach is highly appropriate given that a contemporary phenomenon is being studied within a real-life context (Yin 2009). More specifically within the case study approach, observation is used to provide information on travel patterns as this can capture all movements within a specified time period if designed appropriately, unlike a survey which is subject to non-response bias (Bryman 2001). However, a travel survey of residents is still desirable as information on travel characteristics, attitudes and preferences were being sought, data which is not possible to collect solely through observation. While interviews could capture this information, the number of residents living at the case and control sites would mean that information from only a very small proportion of residents could be realistically attained, thereby limiting the ability to make any generalisations from the data that is collected (Hennink et al. 2011). Finally, secondary data analysis is an appropriate research method given that published data, as a secondary data source, is being used to provide a supplementary comparison to the observational counts.

Further detail on the aim, context, method and results associated with this research component is provided in Chap. 8.

4.3.5 Research Component 5: Application and Integration of Implementation and Planning Enforcement Theories

In addressing research objective 4, this research component takes the findings from components 1–4 and views these through the lens of both implementation theory and planning enforcement theory. This process assists in identifying and assessing opportunities to enhance the implementation (and subsequent effectiveness) of travel plans for new residential developments.

In addition, an integrated theory of implementation and enforcement is developed to guide future travel planning practice for new residential developments. The research findings are assessed in terms of the extent to which they support the integrated theory.

Further detail on the method and results associated with this research component is provided in Chap. 9.

4.4 Limitations of the Research Approach

While the research approach is capable of addressing the research objectives, a number of limitations need to be acknowledged.

Firstly, the online survey of councils was limited to the state of Victoria. This is despite the lack of knowledge concerning the scale of travel planning for new developments in all jurisdictions of Australia (see Table 4.2). Furthermore, the ability to generalise the Victorian findings to other jurisdictions of Australia will be limited given that regulation of land use planning is the responsibility of individual states and territories who each have independent planning systems in place (Department of Infrastructure and Transport 2011). However, resource constraints prevented the survey from being expanded to all councils in Australia as an intensive period of telephone contact was initially required to determine an appropriate representative in each council who could complete the survey, and to then seek their commitment and agreement to complete the survey. In addition, subsequent components of the research were conducted in Victoria, so it was important that efforts were directed to this geographical area rather than attempting to explore all jurisdictions, potentially in less depth.

Secondly, while a total of 30 industry representatives were interviewed, the sample was spread across a range of organisation types. This resulted in only three property developers and three property managers being interviewed as part of the sample. The ability to generalise the interview findings is therefore limited. However, the intent of the interviews was not to provide generalisations, but rather to present a set of actor perspectives on travel planning for new residential developments, particularly aspects relating to implementation.

Thirdly, while a total of 29 travel plans were used for the desktop assessment, these were requested from consultants and government agency representatives who may have been biased in selecting the travel plans they provided. Furthermore, the assessment framework used was developed solely by the researcher. However, this framework was based on a synthesis of best practice elements from the literature. In addition, the travel plans were initially assessed solely by the researcher. This limitation was overcome by requesting a number of travel planning practitioners to apply the framework to a subset of the travel plans, with inter-variability in scores being reported.

Finally, a number of limitations were inherent in the research component concerned with the case studies of new residential developments. Due to the need to collect data at each site, only four residential developments with travel plans could be evaluated within the resources available, particularly given that data collection at matching control sites was also required. Clearly, more sites would help to establish a stronger evidence base concerning the effectiveness of travel plans for new residential developments. Furthermore, a lack of robust monitoring data for residential sites with travel plans (as identified in Chap. 2) meant that a meta-analysis was not possible, despite this being highly desirable.

At each of the case study and control sites, a travel survey of residents was conducted to assess the extent of self-selection. A key limitation was that survey material could only be delivered to residential mail boxes as personal contact (e.g. door knocking) was not permitted by the property managers. Not only did this prevent any interviews or focus groups being conducted with residents (which would have helped to provide additional insight into their perspectives on the travel planning process), it also resulted in a relatively low survey response rate of 14 %. However, this level of survey response was consistent with response rates of 11–20 % achieved in other similar studies (Lee et al. 2014).

Further detail on the methodological issues and limitations associated with each research component is provided throughout Chaps. 5–9.

4.5 Conclusion

The aim of this chapter is to describe the approach for addressing the research gaps and opportunities. In doing so, it outlined five key research components and their alignment to the research objectives. The research components include: an online survey of Victorian councils, interviews with industry representatives, a desktop assessment of travel plans, case studies of new residential developments, and the application and integration of implementation and planning enforcement theories. The use of a mixed-methods approach strengthens the validity of the research through enabling triangulation of the findings. The approach is particularly relevant given the diverse set of research objectives which cannot be achieved using a single method.

A number of limitations inherent in the research approach have been acknowledged. While best efforts have been made to overcome these, opportunities exist to address outstanding limitations in the future. This is covered in Chap. 10 through a discussion of future research directions.

The next chapter (Chap. 5) details the first set of research results. It presents the findings from the online survey of councils to gauge the scale of travel planning practice for new developments in Victoria (research objective 1). Chapter 6 then describes the interview findings which provide an appreciation for the perspectives of industry actors involved in travel planning for new residential developments (research objective 2). This is followed by Chaps. 7 and 8 which consider the quality and effectiveness of travel plans for new residential developments (research objective 3). Through the application and integration of implementation and planning enforcement theories, Chap. 9 identifies and assesses opportunities for enhancing the implementation (and subsequent effectiveness) of travel plans for new residential developments (research objective 4).

References

Bryman, A. (2001). *Social research methods*. Oxford, UK: Oxford University Press.

Department of Infrastructure and Transport. (2011). *Our cities, our future: A national urban policy for a productive, sustainable and liveable future*, Canberra, Australia.

Greenfield, T. (2002). *Research methods for postgraduates* (2nd ed.). London, UK: Arnold.

Hall, R. (2008). *Applied social research: Planning, designing and conducting real-world research*. Victoria, Australia: Palgrave Macmillan.

Hennink, M., Hutter, I., & Bailey, A. (2011). *Qualitative research methods*. UK: SAGE Publications Ltd.

Lee, J. S., Zegras, C., Ben-Joseph, E., & Park, S. (2014). Does urban living influence baby boomer' travel behaviour? *Journal of Transport Geography, 35*, 21–29.

Mack, N., Woodsong, C., MacQueen, K., Guest, G., & Namey, E. (2005). *Qualitative research methods: A data collector's field guide*. North Carolina, USA: Family Health International.

Yin, R. (2009). *Case study research: Design and methods* (4th ed.). California, US: SAGE Publications.

Chapter 5
The Scale of Travel Planning Practice

5.1 Introduction

Chapters 1–4 detailed the background and approach to the research. In doing so, they identified a number of research gaps and opportunities and showed how each of these aligned with the research objectives and components.

This chapter presents the first set of research results by detailing the findings of the online survey of Victorian councils, corresponding to research component 1 Fig. (5.1). Table 5.1 details the research gap, opportunity and objective associated with this research component.

In line with research objective 1, the aim of this chapter is to examine the scale of travel planning practice for new urban developments in Victoria. Key aspects include:

- Understanding the extent to which travel plans have been required by councils
- Identifying the reasons why travel plans have and have not been required
- Identifying mechanisms used to require travel plans
- Assessing the level of monitoring that has taken place to date
- Understanding levels of travel plan familiarity and experience among council staff
- Understanding perceptions of travel plan effectiveness among council staff
- Gauging the likelihood of councils requiring travel plans in the future.

This chapter begins by describing the research method used to examine travel planning practice for new urban developments in Victoria. The results are then presented and compared to the literature described in Chap. 2. The chapter concludes by discussing the implications of the findings for future travel planning practice.

C. De Gruyter, *Travel Plans for New Residential Developments: Insights from Theory and Practice*, Springer Theses,
DOI 10.1007/978-981-10-2092-6_5

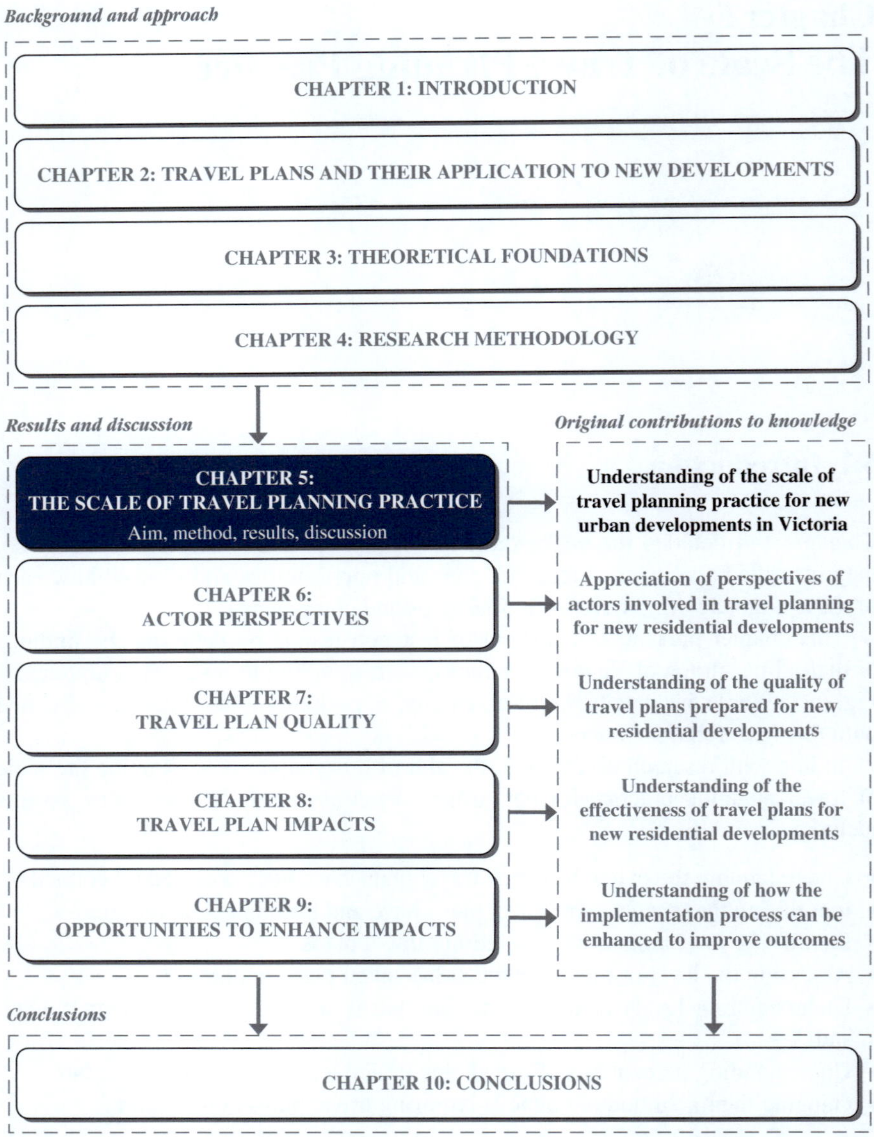

Fig. 5.1 Position of Chap. 5 in the thesis structure

5.2 Research Method

In order to examine the scale of travel planning practice for new developments in Victoria, a self-completion questionnaire was designed and administered online to councils. Anonymity in survey responses was assured so that individual councils could not be identified. Ethics approval was provided by the Monash University

Table 5.1 Research gap, opportunity and objective associated with research component 1

Research gap →	Research opportunity →	Research objective →	Research component
There is no understanding of the scale and associated characteristics of travel planning practice for new developments in Australia	Examine the scale and associated characteristics of travel planning practice for new urban developments, using a case study from the state of Victoria	1. To examine the scale of practice in Victoria, Australia	1. Online survey of Victorian councils

Human Research Ethics Committee (MUHREC) prior to conducting the survey (reference number CF12/1205 – 2012000586).

A survey was considered to be the most appropriate method for achieving the research objective given that answers to 'how much', 'how often' and 'what proportion' type questions were desired (Hennink et al. 2011). This is in contrast to focus groups or interviews which seek to understand beliefs and opinions on particular topics (Mack et al. 2005). The survey was hosted online to simplify the survey administration and data analysis process.

Survey questions covered the extent to which travel plans had been required, reasons for requiring (and not requiring) travel plans, mechanisms used to require travel plans, levels of monitoring undertaken, familiarity and experience with travel plans, perceptions of travel plan effectiveness, and the likelihood of requiring travel plans in the future. A closing question was also included to allow respondents to express any other comments they had about travel plans for new developments. A copy of the survey questionnaire is provided in Appendix C.

All 31 councils in metropolitan Melbourne were targeted for the survey, plus five regional councils (Greater Geelong, Greater Bendigo, Ballarat, Greater Shepparton and Latrobe) which represent the key regional centres across Victoria. This resulted in a total target population of 36 (out of 79) councils, as illustrated in Fig. 5.2. While the remaining regional councils could also have been targeted for the survey, this was not considered appropriate as travel plans are generally not considered in these areas at all due to their predominant rural character. Furthermore, the 36 councils targeted for the survey contain the majority of the Victorian population, at around 84 % (Department of Transport, Planning and Local Infrastructure 2014).

In order to recruit survey participants, an initial telephone call was made to each council to explain the purpose of the survey and to determine an appropriate representative who could complete the survey. Contact with each representative was then made to confirm that they were the most suitable person to participate in the survey. This process took considerable time and often involved multiple telephone calls to each council (more than 10 in some cases) before an appropriate representative could be confirmed and contacted. However, this method was still preferred over other recruitment options (e.g. letter or email) as telephone contact helped to establish commitment from each representative to participate in

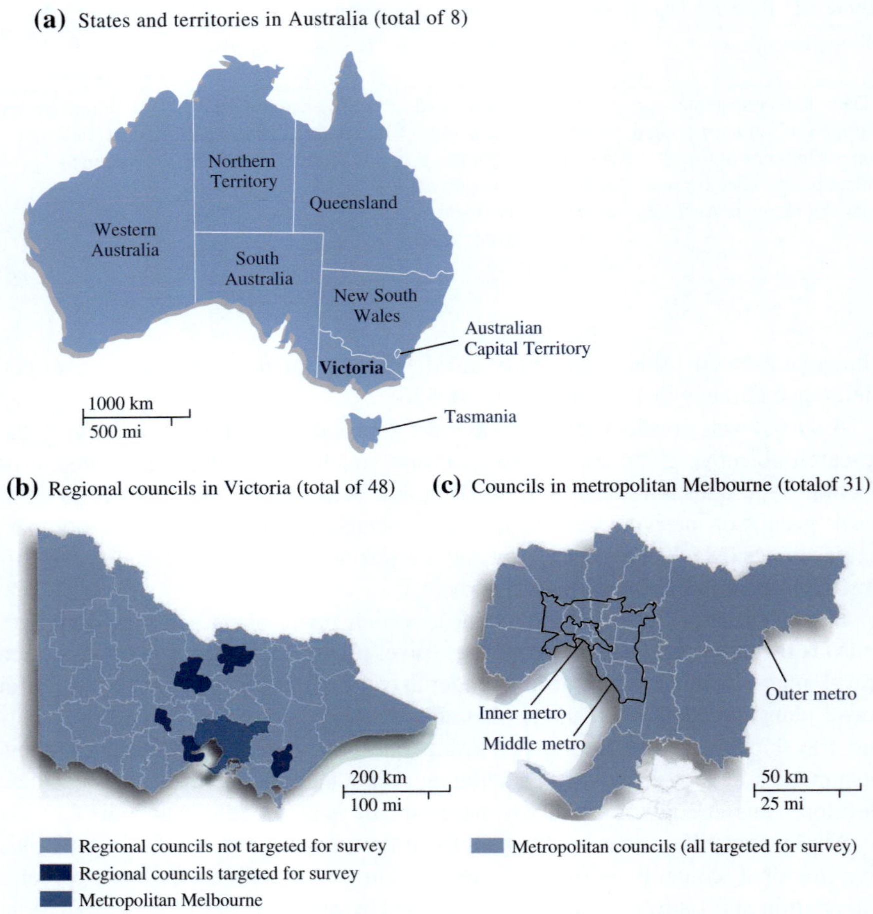

Fig. 5.2 Location of Victorian councils targeted for the survey

the survey. As the survey was administered online, a link could be distributed by email to the selected representative in each council. A reminder email was sent to each representative 1 week later to ensure completion of the survey. The level of interest in the survey was relatively high and there were generally no issues in seeking commitment from each representative to complete the survey once contact had been established with them. As a result, a response from all 36 councils was achieved, representing a 100 % response rate.

An important early finding, revealed during the survey recruitment stage, was that there was little consistency in the administrative unit within each council that was responsible for travel plans for new development. Of the 36 councils surveyed, 19 responses (or 53 %) were provided by representatives in transport related roles, 15 responses (42 %) were provided by representatives in planning related roles, and the remaining two responses (5 %) were provided by representatives in other roles (e.g. environmental management). In some cases, the representative sought input from other relevant areas to complete the survey, highlighting that more than one area within particular councils may be involved in the travel planning process for new developments.

Following completion of the survey, the results were analysed using descriptive statistics. The data was also cross-classified and statistical tests were used to gain insight into the extent to which responses varied according to different segments of the survey population.

5.3 Results

This section presents the results of the survey, in line with each of the survey questions that were asked of council representatives.

5.3.1 Requirements for Travel Plans

Table 5.2 indicates the number of Victorian councils that have previously required a travel plan. Around 80 % of inner and middle metropolitan councils have required a travel plan before, with this figure decreasing to around 20 % for outer metropolitan and regional councils. Overall, 18 out of the 36 councils surveyed (50 %) had previously required a travel plan.

Table 5.3 shows the number of travel plans required by Victorian councils between 2010 and 2012. As can be seen, inner and middle metropolitan councils required more travel plans than outer metropolitan and regional councils. This may be due to higher rates of greenfield development occurring in outer areas compared to inner and middle suburbs where higher density infill development and greater transport issues are experienced.

Half of the councils who had required a travel plan before (equivalent to 25 % of the total sample) had required more than five each during 2010–12. Based on the data presented in Table 5.3, it is estimated that around 100 travel plans were required by Victorian councils during 2010–12.

Table 5.2 Requirements for travel plans for new developments by Victorian councils

Status	Number of councils by location				Total
	Inner metro	Middle metro	Outer metro	Regional	
Required	4 (80 %)	10 (83 %)	3 (21 %)	1 (20 %)	18 (50 %)
Never required	–	2 (17 %)	6 (43 %)	2 (40 %)	10 (28 %)
Unsure	1 (20 %)	–	5 (36 %)	2 (40 %)	8 (22 %)
Total	5 (100 %)	12 (100 %)	14 (100 %)	5 (100 %)	36 (100 %)

Table 5.3 Number of travel plans required by Victorian councils during 2010–12

Number of travel plans required	Number of councils by location				Total
	Inner metro	Middle metro	Outer metro	Regional	
1–2	–	2 (20 %)	1 (33 %)	–	3 (17 %)
3–5	–	–	1 (33 %)	1 (100 %)	2 (11 %)
6–10	2 (50 %)	2 (20 %)	1 (33 %)	–	5 (28 %)
More than 10	1 (25 %)	3 (30 %)	–	–	4 (22 %)
Unsure	1 (25 %)	3 (30 %)	–	–	4 (22 %)
Total	4 (100 %)	10 (100 %)	3 (100 %)	1 (100 %)	18 (100 %)

5.3.2 Reasons for Requiring Travel Plans

Figure 5.3 shows that offsetting the impact of providing reduced car parking was the most common reason for requiring a travel plan for a new development. Reducing car parking has clear benefits for a developer through reduced costs which also acts as an important 'stick' in managing car use as part of the travel plan developed for the site.

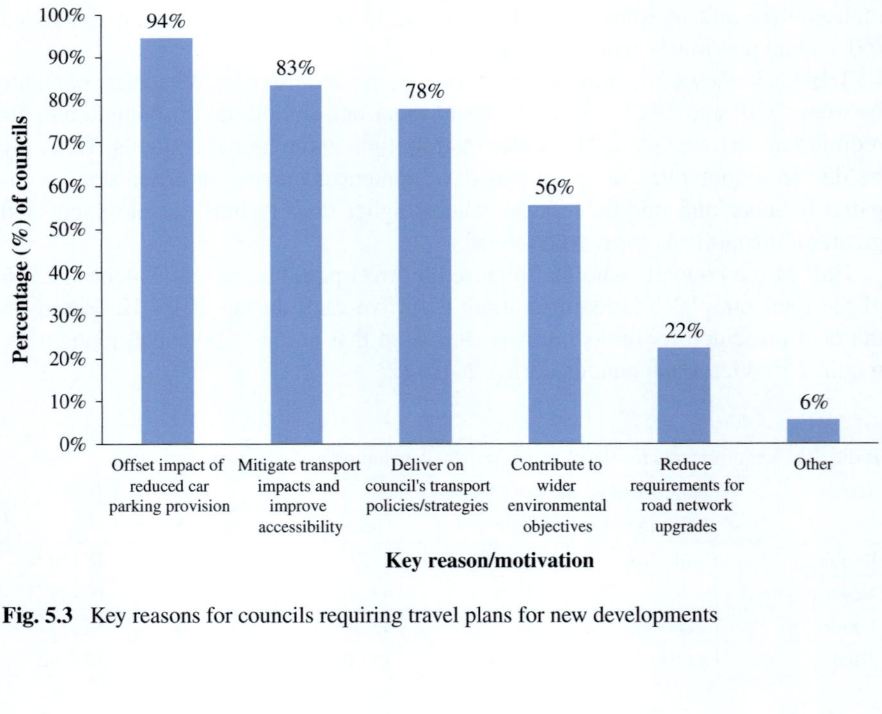

Fig. 5.3 Key reasons for councils requiring travel plans for new developments

5.3.3 Reasons for Not Requiring Travel Plans

The survey results indicated that around one-third of the councils (10 out of the 36 surveyed) had not required a travel plan before. The most common reason for this (stated by five of the 10 councils) was that they did not consider travel plans to be effective or appropriate for their local area. Other reasons included the lack of any statutory requirement, or an intention to require travel plans in the future.

5.3.4 Mechanisms Used to Require Travel Plans

Figure 5.4 shows that a planning permit condition is the most common mechanism used for requiring travel plans for new developments in Victoria. As respondents could indicate more than one mechanism they have used, verbal negotiation (reported by more than one-third of respondents) may have been used in conjunction with other mechanisms. This is because it is unlikely that verbal negotiation alone would be particularly effective in ensuring a travel plan is developed and implemented. Around one-third of respondents (28 %) indicated the use of formal agreements by their council to require travel plans for new developments.

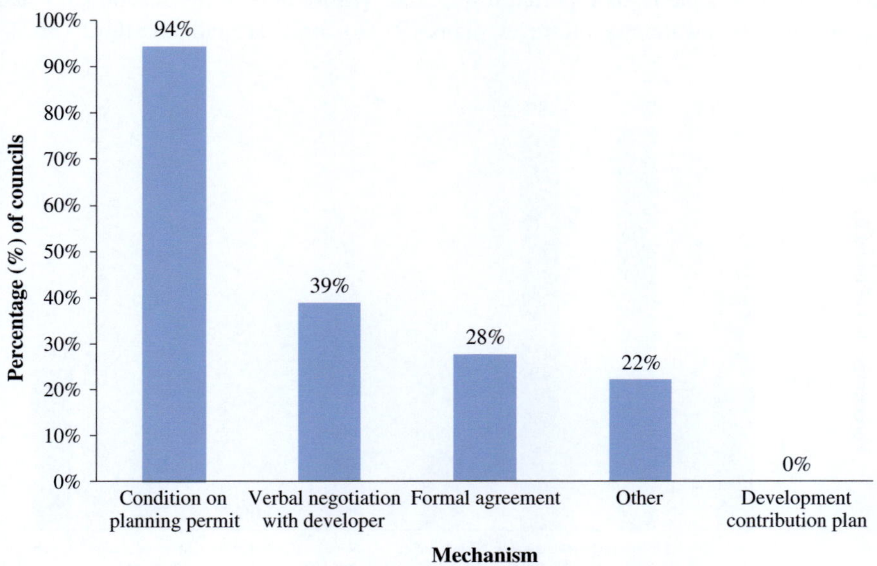

Fig. 5.4 Mechanisms used by councils to require travel plans for new developments

5.3.5 Monitoring of Travel Plans

Figure 5.5 indicates the level of travel plan monitoring undertaken by Victorian councils to date. Around 80 % of councils indicated that they had not undertaken any monitoring of the travel plans they had required. A lack of council resources was a key reason cited for the relatively low rate of monitoring:

> The monitoring of travel plans is complex and it will require more resources from councils to follow up the results in time [Response from outer metropolitan council representative].

However, it was also noted that councils are not necessarily averse to monitoring travel plans, but would only do so if a particular issue arose:

> It is unlikely that we would monitor the plan to ensure it is being implemented, however if a complaint was received in relation to the development and traffic etc., it is likely planning enforcement would ensure that all conditions on the permit (including the green travel plan) are being implemented [Response from middle metropolitan council representative].

5.3.6 Familiarity and Experience with Travel Plans

Council representatives were asked to state their level of familiarity and experience with travel plans, as reported in Fig. 5.6. While most of the respondents had some level of awareness of travel plans (91 %), only around one-third (36 %)

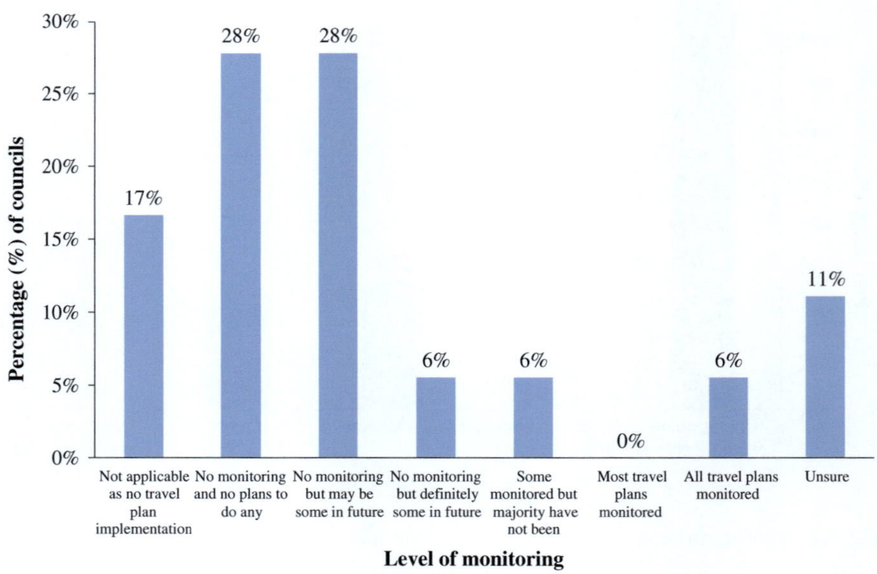

Fig. 5.5 Level of travel plan monitoring among councils

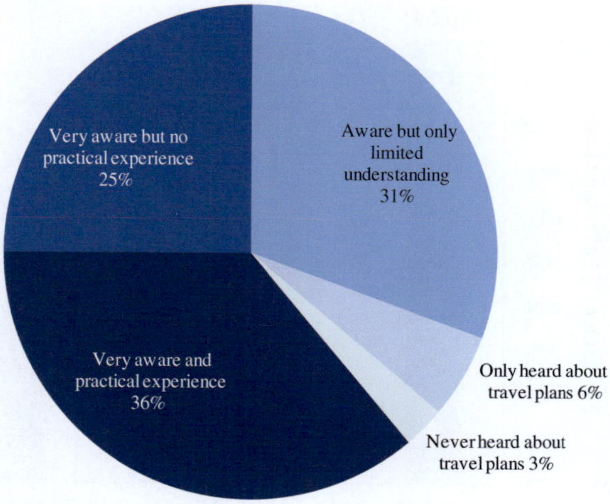

Fig. 5.6 Level of respondent familiarity and experience with travel plans

indicated they had practical experience in using them. Practical experience with using travel plans generally declined with distance from the Melbourne Central Business District (CBD), with inner metropolitan councils reporting the highest proportion of respondents with practical experience (80 %), followed by middle metropolitan councils (42 %), regional councils (20 %) and outer metropolitan councils (14 %).

5.3.7 Perceived Effectiveness of Travel Plans

Council representatives were asked to indicate on a five point likert scale, the extent to which they agreed that a number of mechanisms, including travel plans, are effective in managing transport access at new developments. The proportion that either agreed or strongly agreed that each mechanism is effective is shown in Fig. 5.7. As can be seen, travel plans ranked lowest of the mechanisms presented with only one in two respondents (50 %) regarding them as effective.

Table 5.4 shows the extent to which respondents agreed that travel plans are effective and cross-classifies this by their familiarity and experience with travel plans. In total, 64 % of those familiar and experienced with travel plans agreed (or strongly agreed) that travel plans are effective, compared to only 29 % of those with a limited understanding of how travel plans work. A z-test for the difference between proportions showed that there was a statistically significant difference to suggest that those familiar and experienced with travel plans are more likely to agree that they are effective in managing transport access to new developments ($p = 0.04$).

Table 5.4 also shows that around one-third of respondents (36 %) felt 'neutral' towards the effectiveness of travel plans. However, when cross-classified, it can

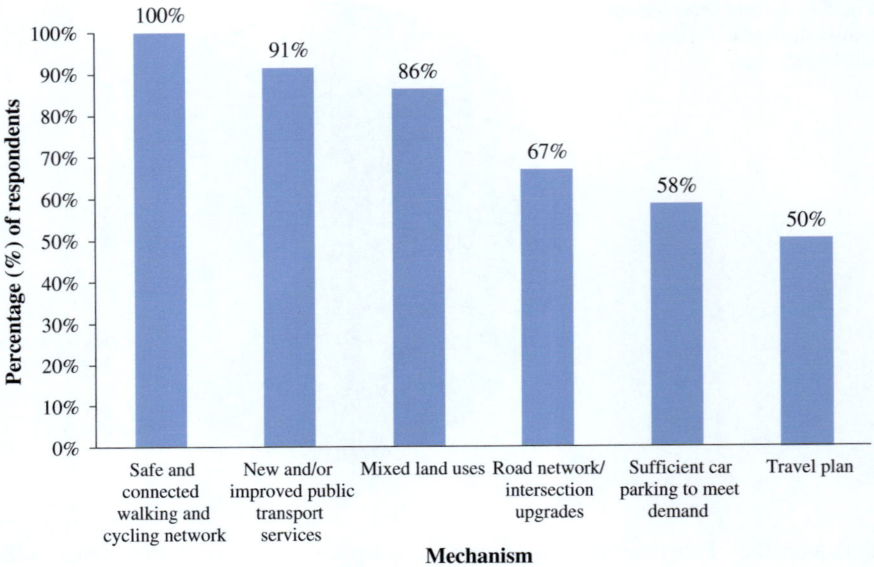

Fig. 5.7 Mechanisms considered effective for managing transport access at new developments

Table 5.4 Extent to which respondents agreed that travel plans are effective in managing transport access for new developments (cross-classified by their familiarity and experience)

Extent of agreement that travel plans are effective	Respondent familiarity and experience with travel plans		Total
	Familiar and experienced	Not familiar or experienced	
Agree or strongly agree	14 (64 %)	4 (29 %)	18 (50 %)
Neutral	6 (27 %)	7 (50 %)	13 (36 %)
Disagree or strongly disagree	2 (9 %)	3 (21 %)	5 (14 %)
Total	22 (100 %)	14 (100 %)	36 (100 %)

be seen that only 27 % of those familiar and experienced in using travel plans felt 'neutral' towards their effectiveness, compared with 50 % of those with a limited understanding of how travel plans work. Despite this finding, a z-test for the difference between proportions showed there was no statistical significance to suggest that those not familiar and experienced with travel plans were more uncertain about their effectiveness ($p = 0.17$).

In addition, perceived effectiveness of travel plans was also cross-classified by council location and the number of travel plans they had required for new developments. However, no clear pattern or differences emerged from these additional analyses.

Finally, it should be noted that uncertainty about travel plan effectiveness did not necessarily correspond with disinterest in the concept:

> We are very interested in exploring the use of travel plans for new developments, however we have not come across any cases where they have been proven to be effective… [Response from outer metropolitan council representative].

5.3.8 Future Likelihood of Requiring Travel Plans

Figure 5.8 shows that around half of the councils (51 %) were likely or highly likely to require a travel plan for a new development in the next 12 months, which is similar to the proportion that have required them previously (50 %). The proportion of councils that were likely or highly likely to require a travel plan in the next 12 months also declined with distance from the Melbourne CBD. Inner metropolitan councils were most likely (80 %), followed by middle metropolitan councils (75 %), regional councils (50 %) and outer metropolitan councils (21 %).

5.3.9 Other Key Issues

A number of other key issues were highlighted by councils as part of the survey. These included the lack of any state planning policy that is supportive of travel plans and concerns about the effectiveness of travel plans:

Fig. 5.8 Likelihood of Victorian councils requiring a travel plan in the next 12 months

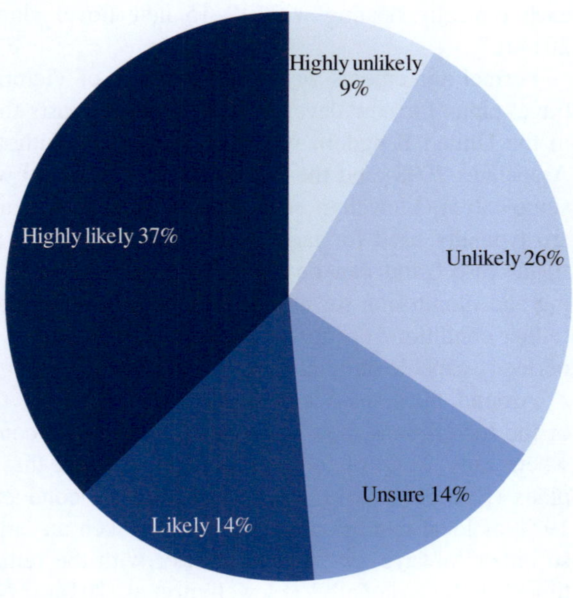

> The town planning process is very focussed on vehicular movements with little consideration towards other modes (apart from broad motherhood statements in the State Planning Policy Framework) therefore it is difficult to liaise with some developments in requesting extra info such as travel plans [Response from outer metropolitan council representative].

> Travel plans are not effective when implemented by a statutory requirement, like many management plans [Response from middle metropolitan council representative].

5.4 Discussion

This section compares the survey findings to the literature described earlier in Chap. 2 and discusses the implications for future travel planning practice.

It is insightful to compare the results to a previous survey conducted of 388 local authorities in the United Kingdom (Steer Davies Gleave 2001). The survey was undertaken in 2000 to assess the take-up and effectiveness of travel plans, and included a component on travel plans and development control. The results of this survey showed that 58 % of local authorities had required a travel plan for a new development, compared to 50 % for the Victorian council survey. However, the percentage today in the United Kingdom is likely to be much higher due to the increased focus on travel plans for new developments and the ongoing presence of a supportive policy framework (Rye et al. 2011b). A subsequent survey of local authorities in the United Kingdom in 2007 suggested a three-fold increase in travel plans between 2001 and 2006, although it was also recognised that most local authorities currently secure less than 10 travel plans each year (Addison & Associates 2008). A more recent survey conducted in Scotland in 2013 showed that every local authority now requires travel plans for new developments, with each typically dealing with 12–15 new travel plans per year (Llewellyn et al. 2014a).

Formal agreements were used by 28 % of Victorian urban councils to require travel plans for new developments. This contrasts the finding of the 2007 survey in the United Kingdom where 61 % of local authorities used them (Addison & Associates 2008) and the 2013 survey in Scotland where 86 % had used them to some extent (Llewellyn et al. 2014a). Formal agreements in the United Kingdom are typically used for larger developments as they are considered to have more 'legal force' and can be used to secure payments associated with implementing and monitoring travel plans. However, as argued by Harrison (2003, p. 401), 'either conditions or [agreements] concerning travel plans if properly drafted, are perfectly capable of being lawful and enforceable.'

Around 80 % of Victorian councils indicated they had not monitored any of the travel plans they had required. This is in contrast to the United Kingdom where only 21 % of local authorities reported that they did not monitor travel plans (Rye et al. 2011a). The 2013 survey conducted in Scotland showed that 19 % of local authorities had not undertaken any monitoring, yet only 25 % did so either 'always' or 'in most cases', with the remainder doing so only 'sometimes' or 'occasionally' (Llewellyn et al. 2014a). Nevertheless, a very different pattern of travel plan monitoring exists between Victoria and the United Kingdom. Difficulties with monitoring and enforcing travel plans have still been reported in the United Kingdom (Llewellyn et al. 2014b; Rye et al. 2011a), along with the United States (Seggerman and Hendricks 2005) and the Australian state of New South Wales (Wynne 2013). Findings from the 2007 survey in the United Kingdom echo comments made by Victorian councils regarding limited resources:

> Resourcing (or the lack of it) of the monitoring, penalties, sanctions and incentives processes was seen by many authorities as a reason for not including them within travel plans as they have no resources to follow this through (Addison & Associates 2008, p. 71).

The lack of any state planning policy that is supportive of travel plans was also identified as a key issue with current practice in Victoria. In the United Kingdom, a strong policy context was the most cited 'assisting' factor by councils in securing travel plans for new developments (Addison & Associates 2008). A supportive policy framework is also considered to be important in other European countries (Rye et al. 2011b).

The level of practical experience with using travel plans was not particularly high among council staff in Victoria, yet those with experience were more likely to perceive travel plans as effective. Addison & Associates (2008, p. 78) note that a 'lack of knowledge hindered the effectiveness of the travel plan work' in the United Kingdom and that 'training was seen as much needed'.

Finally, it is worth noting again that around half of the councils surveyed stated that they were likely to require a travel plan for a new development in the future. This clearly indicates an area of continued growth in the area of transport and land use planning, demonstrating the importance of identifying opportunities to enhance travel planning practice where possible.

The survey findings suggest a number of opportunities for enhancing future practice. Examples include: changing monitoring and enforcement practices, introducing more supportive planning policies, and developing relevant guidance material and training programs. These, along with other opportunities, are discussed further in Chap. 9.

5.5 Conclusion

The aim of this chapter is to examine the scale of travel planning practice for new developments in Victoria. In doing so, it showed that half of the councils had previously required a travel plan for a new development, primarily to offset the impact of providing less car parking. It was estimated that around 100 travel plans were required by Victorian councils during 2010–12 alone. However, little monitoring of travel plans has taken place to date which, among others, was identified as a key issue by councils.

From the results presented, it appears that Victoria is still at a somewhat embryonic stage with respect to travel planning for new developments, which is perhaps analogous to the United Kingdom's position 10 years ago. With a better understanding of travel planning practice for new developments in Victoria, a number of opportunities can be identified to move towards a more effective approach. These opportunities are discussed in Chap. 9.

The survey findings presented in this chapter have laid an important foundation for future chapters of this thesis. They have shown that while a modest level of travel planning activity for new developments has occurred in Victoria, various issues have been experienced that warrant further investigation. Building upon the survey findings, subsequent chapters of this thesis focus specifically on travel plans for new residential developments. The next chapter presents the results from a series of interviews aimed at developing an appreciation for the perspectives of actors involved in their application, with a particular focus on implementation.

References

Addison & Associates. (2008). *Delivering travel plans through the planning process—Research report*. London, UK: Department for Transport and Communities and Local Government.

Department of Transport, Planning and Local Infrastructure. 2014. *Victoria in future 2014 data tables*. Australia: Victoria. http://www.dpcd.vic.gov.au/.

De Gruyter, C., Rose, G., & Currie G (2014). 'Securing travel plans through the planning approvals process: A case study of practice from Victoria, Australia', *Cities, 41*, part A, 114–122.

Harrison, J. (2003). Travel plans and the planning system. *Journal of Planning and Environment Law*, no. April 2003, 397–403.

Hennink, M., Hutter, I., & Bailey, A. (2011). *Qualitative research methods*. UK: SAGE Publications Ltd.

Llewellyn, R., Paton, D., & Tricker, R., (2014a) *After the downturn: where now for travel plans in Scotland?* Paper presented to Scottish Transport Applications & Research (STAR) conference, Glasgow, Scotland.

Llewellyn, R., Tricker, R., & Paton, D. (2014b) Travel plans: A critical comparison of the application of land use planning processes in England and Scotland, *Transport*, no. i First, 1–13.

Mack, N., Woodsong, C., MacQueen, K., Guest, G., & Namey, E. (2005). *Qualitative research methods: A data collector's field guide*. North Carolina, USA: Family Health International.

Rye, T., Green, C., Young, E., & Ison, S. (2011a). Using the land-use planning process to secure travel plans: An assessment of progress in England to date. *Journal of Transport Geography, 19*(2), 235–243.

Rye, T., Welsch, J., Plevnik, A., & de Tomassi, R. (2011b). First steps towards cross-national transfer in integrating mobility management and land use planning in the EU and Switzerland. *Transport Policy, 18*, 533–543.

Seggerman, K., & Hendricks, S. (2005). *Incorporating TDM into the land development process*. State of Florida: Department of Transportation.

Steer Davies Gleave. (2001). *Take-up and effectiveness of travel plans and travel awareness campaigns: Final report*. London, UK: Department of the Environment, Transport and the Regions.

Wynne, L. (2013). *Travel planning and the land-use planning system: Understanding the effectiveness of travel planning requirements in NSW Development Control Plans*, Unpublished Masters Thesis, University of New South Wales, Australia.

Chapter 6
Actor Perspectives

6.1 Introduction

The previous chapter presented the first set of research results to provide an understanding of the scale of travel planning practice for new urban developments in Victoria, Australia. In doing so, it identified a range of issues with the process of requiring travel plans.

This chapter, as positioned in Fig. 6.1, presents the second set of research results by detailing the findings from interviews conducted with industry representatives (research component 2). Table 6.1 details the research gap, opportunity and objective associated with this research component.

In line with research objective 2, the aim of this chapter is to gain an appreciation for the perspectives of industry actors involved in travel planning for new residential developments. Key aspects include:

- Perceived benefits and potential disadvantages
- Extent of industry involvement and interactions among stakeholders
- Implementation challenges and potential solutions
- Future expectations.

This chapter begins by describing the research method used to gain an appreciation for the perspectives of industry actors. The results are then presented and compared to the literature described in Chap. 2. The chapter concludes by discussing the implications of the findings for future travel planning practice.

C. De Gruyter, *Travel Plans for New Residential Developments: Insights from Theory and Practice*, Springer Theses,
DOI 10.1007/978-981-10-2092-6_6

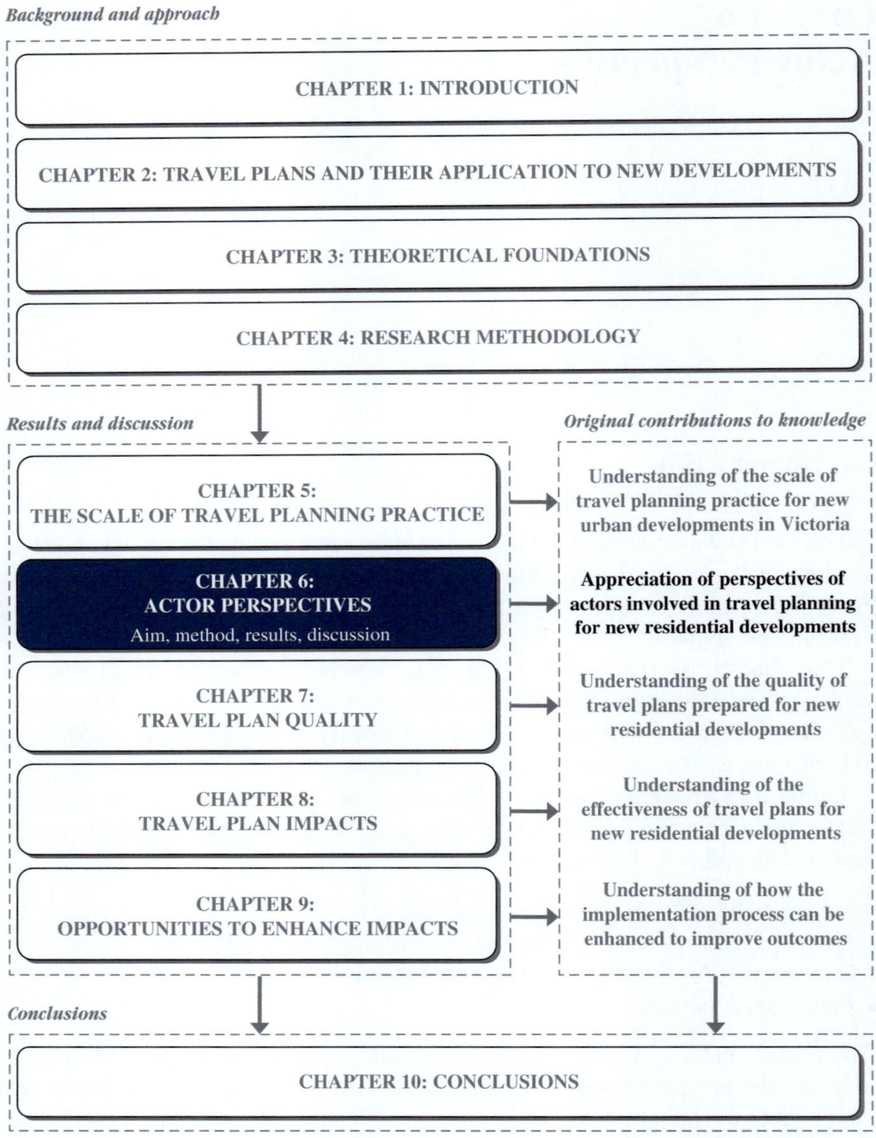

Fig. 6.1 Position of Chap. 6 in the thesis structure

6.2 Research Method

Interviews were considered to be the most suitable approach for achieving the research objective as perceptions, beliefs and experiences were being sought from industry representatives. This is in contrast to 'how much' or 'what proportion' type questions where a survey is generally more suitable (Richardson et al. 1995).

Table 6.1 Research gap, opportunity and objective associated with research component 2

Research gap →	Research opportunity →	Research objective →	Research component
No research has specifically explored the perspectives of the different actors involved in travel planning for new residential developments	Develop an appreciation for the perspectives of actors involved in travel planning for new residential developments, particularly aspects relating to implementation	2. To gain an appreciation for the perspectives of industry actors involved in their application	2. Interviews with industry representatives

A semi-structured interview approach was chosen to retain flexibility, thereby allowing greater emphasis to be placed on some of the interview topics as required. The approach chosen was found to be highly appropriate given the diversity of roles among those that were interviewed. Ethics approval was provided by the Monash University Human Research Ethics Committee (MUHREC) prior to conducting the interviews (reference number CF12/1205—2012000586).

In identifying the target group for the interviews, all types of actors from industry that have been involved, or may be involved in the future, in travel planning for new residential developments were considered. This included representatives from state and local government, property developers, property managers, and consultants. While property developers and managers are not traditionally seen as a core part of the transport industry, they were included given their current and potential future involvement with travel plans. Existing industry contacts were first used to recruit participants, followed by the adoption of a 'snowballing' technique where participants that had already been interviewed were asked to suggest other representatives that may be suitable for interviewing (Hennink et al. 2011; Mack et al. 2005).

As multiple interviews are needed to ensure a range of views are heard on a given topic (Hennink et al. 2011), it was ensured that at least three interviews were undertaken with each type of organisation. In total, 20 semi-structured interviews were conducted with 30 industry representatives (some interviews involved more than one participant). The potential for the 'snowballing' recruitment technique to introduce bias to the sample was acknowledged (Hennink et al. 2011) and therefore adopted in only five out of the 20 interviews. As travel planning for new residential developments is more established in England than in Australia, three of the interviews were held with travel planning representatives from England via telephone (or Skype). The remaining 17 interviews were held in person with industry representatives from Australia.

Table 6.2 details the interview participants and their general experience with travel plans. Most of the participants worked in either a transport planning/engineering or planning related role, with two of the participants (C6 and C7) working solely in travel planning roles. General experience with travel plans varied among interview participants, with a greater level of experience evident among those working in state government agencies and consultancies.

Table 6.2 Interview participants and their general experience with travel plans

Organisation type and [ID]	Number of interviews	Number of participants	Role of participant in organisation and [ID]	General experience with travel plans[a]
Local Government [LG]	4	9	Environmental Sustainable Development [LG1]	Moderate
			Sustainable transport [LG2]	High
			Transport planning [LG3]	High
			Strategic planning [LG4]	Moderate
			Transport and traffic engineering [LG5]	Low
			Strategic planning [LG6]	Moderate
			Statutory planning [LG7]	Low
			Planning [LG8]	None
			Transport engineering [LG9]	None
State Government [SG]	4	7	Statutory planning [SG1]	High
			Transport planning [SG2]	Moderate
			Structure planning [SG3]	None
			Structure planning [SG4]	Low
			Planning tribunal member [SG5]	High
			Planning tribunal member [SG6]	High
			Business engagement (transport) [SG7]	High
Property Developer [PD]	3	3	Project director [PD1]	Moderate
			Director [PD2]	Low
			Development director [PD3]	High

(continued)

Table 6.2 (continued)

Organisation type and [ID]	Number of interviews	Number of participants	Role of participant in organisation and [ID]	General experience with travel plans[a]
Property Manager [PM]	3	3	Strata management [PM1]	None
			Owners corporation management [PM2]	None
			Owners corporation management [PM3]	Low
Consultant [C]	5	7	Planning [C1]	Moderate
			Transport planning [C2]	High
			Project management (transport) [C3]	High
			Project management (transport) [C4]	High
			Transport and traffic engineering [C5]	High
			Travel planning [C6]	High
			Travel planning [C7]	High
Other [O]	1	1	Transport planning [O1]	Low
Total	20	30		None = 5 Low = 6 Moderate = 6 High = 13

[a]General experience with travel plans among interview participants was defined as follows
None no experience with travel plans before
Low some awareness but limited involvement (less than 2 years experience)
Moderate good understanding of travel planning (2–5 years experience)
High more than 5 years experience or significant exposure to travel plans for both existing and new developments

While the requirement to conduct at least three interviews with each type of organisation was met, extensive efforts were required to reach this minimum for property developers and managers. This was because they did not see transport as a core function of their business, particularly given the relative novelty of the travel planning concept for the residential development industry. This finding has implications for travel planning at new residential developments given that property developers and managers are typically the closest to residents of all industry actors and therefore arguably have the greatest influence on the relative success of the travel plan.

Given the sample of 30 interview participants and the diversity of organisation types, coupled with the qualitative nature of the research approach, the ability to generalise the findings is limited (Bryman 2001). However, the intent of the interviews was not to provide generalisations, but rather to present a set of actor perspectives on travel planning for new residential developments, particularly aspects relating to implementation.

The interview questions are shown in Table 6.3. The first two questions focused on the background of interview participants which helped to build rapport and set the context for the remaining questions. The interview then focused specifically on travel plans at new residential developments through a discussion of their benefits and potential disadvantages, current involvement, interactions with other organisations/actors, implementation challenges and potential solutions, and future expectations. A closing question helped participants to 'fade out' from the interview and provide an opportunity to discuss anything that was not covered.

To reduce the potential for reflexivity bias [in which participants give an answer they think the interviewer wants to hear (Yin 2009)], anonymity of responses was assured both from an individual and organisational perspective and the importance of responding honestly was reiterated to participants. This enabled participants to talk openly about travel plans, thereby providing an accurate and honest set of perspectives.

Table 6.3 Interview questions

Background
1. Firstly, can you tell me about your position and its role in the organisation? 2. Have you had any general experience with travel plans before?
Benefits and potential disadvantages
3. What do you see are the benefits of travel plans at new residential developments? 4. What do you see are the potential disadvantages of travel plans at new residential developments?
Current involvement
5. Can you describe any specific residential developments you've been involved with that have required a travel plan? 6. Are there other organisations that you've dealt with before through your involvement in travel plans at new residential developments? If so, to what extent have you turned to them for assistance on matters relating to travel plans?
Implementation challenges
7. What do you think are some of the challenges with implementing travel plans at new residential developments? 8. What are some of the ways that these implementation challenges could be overcome?
Future expectations
9. What do you think the future holds for travel plans at new residential developments? 10. Do you have any other comments you'd like to make about travel plans in general?

On average, each interview lasted for one hour, although these ranged from 30 to 90 minutes. All interview participants, except one, agreed to have their interview audio-taped. For the interview that was not recorded, additional notes were taken during and after the interview which were sufficient for analysis purposes.

Following the conduct of each interview, the audio-recording was transcribed. Key quotes were recorded verbatim, with other responses documented in summary format. In accordance with guidance on qualitative data analysis (Hennink et al. 2011), the interview data was analysed using a set of codes that were developed inductively based on key themes that arose from the interview responses.

6.3 Results

This section presents the interview findings. Industry involvement with travel plans at new residential developments is covered first, followed by their perceived benefits and shortcomings. Interactions between organisations/actors are then presented, along with an overview of key implementation challenges and potential responses. The section concludes with coverage of future expectations and a synthesis of the interview findings.

6.3.1 Industry Involvement with Travel Plans for New Residential Developments

Property managers had a relatively low level of involvement with travel plans at new residential developments (only 1 out of the 3 interview participants [PM3] had any involvement), while most interview participants from local government [LG1–LG7] and consultancies [C2–C7] had been involved in three or more residential developments that had required a travel plan.

Organisational involvement with travel plans for new residential developments spanned a number of areas. As shown in Table 6.4, more than half of the organisations that were interviewed (11 out of 20) had involvement with preparing/developing travel plans for new residential developments, yet only five organisations had involvement in implementation. Furthermore, across all 20 interviews, only two (which were both with consultants) indicated any involvement with monitoring travel plans at new developments.

By their very nature as authorities, local and state government were more likely to only require travel plans. Property developers and consultants, on the other hand, were more likely to be involved in preparing/developing travel plans, and to a lesser extent, implementing them.

Table 6.4 Organisational involvement in travel plans for new residential developments

Area/s of involvement stated by interview participants	No. of interviews where response was given*						Total* (N = 20)
	[LG] (n = 4)	[SG] (n = 4)	[PD] (n = 3)	[PM] (n = 3)	[C] (n = 5)	[O] (n = 1)	
Requiring travel plans	4	3	–	–	–	–	7
Preparing/developing	1	1	3	–	5	1	11
Implementing	–	–	2	1	2	–	5
Monitoring	–	–	–	–	2	–	2
No involvement	–	–	–	2	–	–	2

*More than one response could be given in each interview
[LG] = Local Government; [PM] = Property Manager
[SG] = State Government; [C] = Consultant
[PD] = Property Developer; [O] = Other

6.3.2 Perceived Benefits of Travel Plans for New Residential Developments

The environmental and social benefits of travel plans at new residential developments were the most commonly cited benefits, referred to in 9 out of the 20 interviews, including all 4 interviews with local government.

Reducing the amount of space required for car parking was also commonly cited. Interestingly, only one property developer cited this benefit [PD3], despite the cost advantages that reduced car parking can deliver. However, reduced car parking can also cause unintended parking problems on surrounding streets if appropriate controls are not put in place. This issue was raised by all property managers interviewed.

> Limited car parking, I can see the sustainability and environmental aspect, but in reality, you've got all of these people and where are they going to park their cars? It causes drama and chaos [PM3].

Requiring a travel plan was seen as a way to force developers to consider sustainable transport issues at new residential developments.

> If the council hadn't made us do it, we probably wouldn't have, but it's becoming part of that industry that we're doing it and it's not that big an impost so we can afford it…so I think it works quite well [PD1].

A range of other benefits were stated by interview participants, including the ability to:

- Change residents' attitudes/mindsets towards sustainable transport [LG2, SG2]
- Target residents at a time of change [LG3, SG1, SG7, C6]
- Reduce costs for developers [LG6, PD3, C7]
- Provide a selling feature for developments [PD1, PD3, PM2, C1, C6]

- Achieve a better development outcome through locked-in infrastructure [LG1, LG7, PM1]
- Achieve council's goals [LG1, C6]
- Encourage diversity in the occupation of developments [PD1, PD3, PM2].

6.3.3 Potential Shortcomings of Travel Plans for New Residential Developments

The lack of any follow-up or enforcement was the most commonly cited shortcoming of travel plans at new residential developments, referred to in 9 out of the 20 interviews. Consultants were more likely to cite this disadvantage, with 4 out of 5 interviews with consultants referring to it.

> Even if [the council] asked for loads of stuff in that travel plan, a lot of them won't come back and check you're doing it. As a consultant that delivers travel planning, I'd much rather have councils ringing me up every five minutes saying 'Have you done your survey yet? Have you launched that bike to work scheme? …because then I could go back to the [developer] and say you have to do it. Until they get told that, it's really difficult to convince them they've got to do it [C6].

A lack of adequate transport infrastructure to support travel plans, such as public transport and safe facilities for waking and cycling, was also commonly cited. This was particularly prevalent in discussions relating to greenfield sites in outer metropolitan areas:

> The lack of public transport in outer areas is a very significant constraint to the success of any travel plan [SG2].

The lack of any government policy or legislative requirement for travel plans in Australia was also raised by interview participants [LG2–LG4, LG7, LG9, SG1, SG5, SG6, PD1, PD2, C2–C5, O1]. This included an interview with a property developer [PD1] who commented on the unregulated nature of travel plans required for new developments:

> There's no rules about this, it's very unregulated. It's not clear in terms of what you have to do or why you have to do it [PD1].

The issue of varying quality in travel plan documents was raised in 6 out of the 20 interviews. Local and state government, as reviewers and approvers of travel plans, accounted for 5 out of the 6 interviews that raised this as an issue.

In 5 of the 20 interviews, participants felt that there were no disadvantages associated with travel plans at new residential developments. However, there were also some views expressed that they need to be delivered effectively and should not be relied on as a 'silver bullet' solution.

> I don't see any particular disadvantages. I don't see that they're the panacea of all ills, however [PD2].

A range of other potential disadvantages were raised by interview participants, including:

- Developers paying lip service to travel plans [LG1, LG3, SG7, C3, C6, O1]
- Local context often not considered when requiring travel plans [PD3, PM1, C1, C2, C5–C7]
- Uncertainty surrounding responsibilities for implementation [LG2, LG3, LG5, C2, C7]
- Lack of guidelines on travel plans for new residential developments [LG4, PD1, C4]
- Resource intensive nature of residential travel planning [LG4, PD1]
- Travel plan requirement coming too late in the planning application process [C3, C7]
- General lack of ownership of the travel plan [LG2, SG1, C7].

6.3.4 Interactions Between Organisations/Actors

Interview participants were asked about other organisations they have dealt with through their involvement in travel planning for new residential developments and the extent to which they have turned to them for assistance. Based on responses to this question, Fig. 6.2 provides an overview of the interactions that occur between actors on travel plans for new residential developments. A thicker line and larger

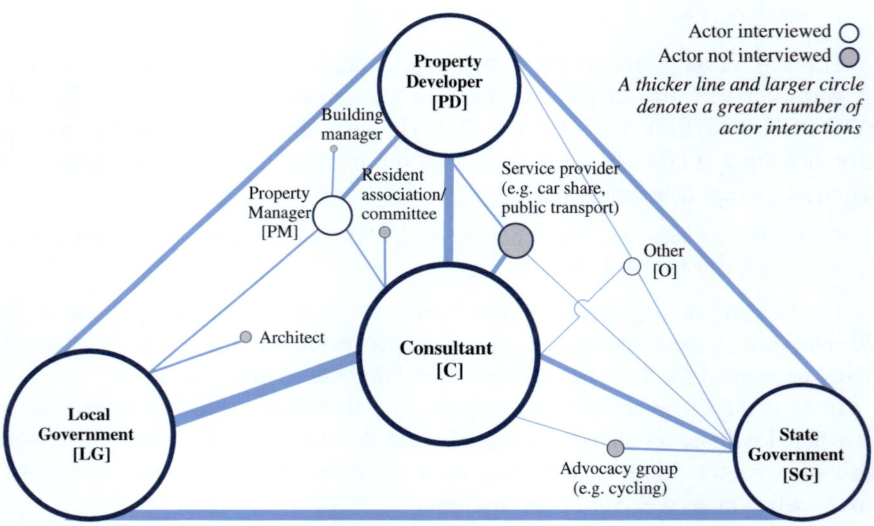

Fig. 6.2 Interactions between actors on travel plans for new residential developments. *Source* Author's synthesis based on responses from interview participants to question 6. Note: *circles* and *lines* drawn to scale and normalised

circle in the figure denotes a greater number of actor interactions. The lines and circles are drawn to scale and normalised to account for differences in the number of interview participants by organisation type. This shows that consultants and local government, and to a slightly lesser extent, property developers and state government, are the key actors currently involved in the process. Consistent with the findings presented earlier in Table 6.4, property managers currently play a relatively minor role in travel planning for new residential developments, despite them arguably having the closest relationship with residents of all industry actors. This may be symptomatic of a lack of travel plan implementation to date.

The interview responses also highlighted other actors involved in the process who were not interviewed. These included architects, building managers, resident associations/committees, service providers and advocacy groups.

It is also noted that other interactions between actors may exist, but were not stated by interview participants. For example, consultants are likely to deal with architects on matters relating to car and bicycle parking facilities. Furthermore, local government may deal with service providers such as car share operators in identifying suitable locations for car sharing vehicles.

6.3.5 Challenges Associated with Implementation

Interview participants raised a number of challenges associated with implementing travel plans at new residential developments. These are illustrated using the word cloud shown in Fig. 6.3. The word cloud illustrates the frequency at which responses were given. If a particular response was given twice as many times as other responses, that response will appear twice as large in the word cloud. As interview participants did not generally express common responses in precisely the same way as each other, the responses contained in the word cloud are not presented verbatim and are instead based on the author's interpretation.

Consistent with the 'potential shortcomings' reported earlier, the lack of any enforcement and/or difficulties associated with enforcement was the most commonly cited implementation challenge, referred to in 8 out of the 20 interviews. This response was most commonly raised by consultants, with 4 out of the 5 consultant

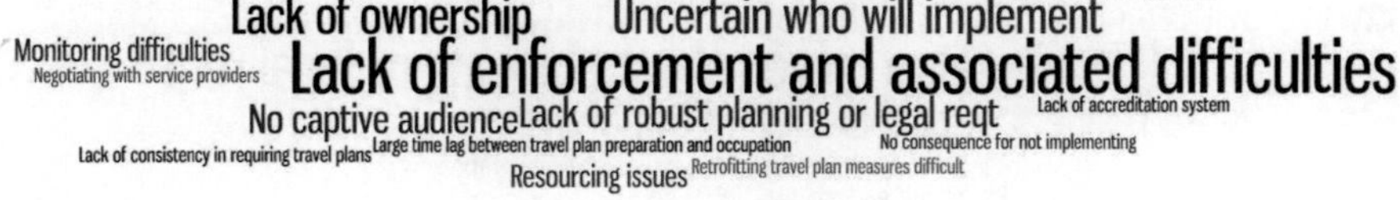

Fig. 6.3 Challenges with implementing travel plans at new residential developments. *Note* phrases contained in the word cloud are based on the author's interpretation of interview responses

interviews referring to it. Difficulties associated with enforcement were generally related to resourcing issues within local government.

> I think this is where the biggest hole is…councils are the ones that are actually requiring these things so they're the logical enforcement body for them, but they clearly don't have the funding or the resources to deal with these things… [C5].

Uncertainty around who will implement the travel plan was also commonly cited. This issue is reflective of a number of other issues raised by participants such the lack of ownership, and the lack of any government regulation or guidance that would typically stipulate implementation responsibilities.

> Whether they actually get carried out when the development is constructed…that's the big question, I don't know…because we don't have involvement from that later stage onwards. We would be advantaged if we had involvement all the way through, through preparation to implementation [C2].

As shown in Fig. 6.3, a range of other implementation challenges were identified, including the lack of a robust planning/legal requirement for residential travel plans, consistent with the potential shortcomings cited earlier.

> There's no state planning policy for travel plans…it's up to local councils to have something in their local policy framework…and it depends on what the focus of the councils are [SG6].

6.3.6 Potential Responses to Implementation Challenges

Interview participants suggested a range of potential responses to the challenges associated with implementing travel plans at new residential developments. These are illustrated in the word cloud shown in Fig. 6.4.

Developing a more robust planning/legal requirement for travel plans at new developments was the most commonly cited response, referred to in 6 out of the 20 interviews.

> The whole regulation thing; there's so much scope for this to really get to a clearer point for people so that it does become a better outcome for everyone, and I think that if that was the case then it would be a real plus moving forward [PD1].

Fig. 6.4 Potential responses to implementation challenges. *Note* phrases contained in the word cloud are based on the author's interpretation of interview responses

However, some caution was expressed in having a requirement that may be too prescriptive or not applicable to all geographical areas:

> If you be too prescriptive, you limit or you almost chop off the ability for innovation…I think it could be detrimental in that way [LG6].

Having the developer fund the implementation of the travel plan was raised as a suggestion in 5 out of the 20 interviews, which included 2 of the 3 property developers [PD1, PD3] that were interviewed.

> I suppose that's one of the key aspects of a green travel plan is ensuring that the developer actually puts his money where his mouth is… [PD3].

Involving the owners corporation and encouraging resident engagement in the process were also raised as suggestions by interview participants [SG6, PD2, C4, C7]. This is particularly relevant given that residents are the target group for residential travel plans. The suggestion of having an independent not-for-profit organisation to handle implementation was also raised:

> I could see that there might be a market out there for that, and developers would like it because if they can just pay a certain amount, the obligation is then established, they walk away…I could see a big advantage to owners corporations where that's one less thing that they have to manage [SG5].

6.3.7 Future Expectations of Travel Plans for New Residential Developments

When asked about the future of travel plans for new residential developments, most participants felt they were either here to stay or will increase in focus (stated in 13 out of the 20 interviews).

> I think they're going to come into play more so moving forward and I suppose if you're going to be reducing parking rates and approving developments with little or no car parking…then you need a fall-back, whether that's public transport, travel plans, or whatever it might be [C2].

However, the view that travel plans are likely to reduce in scale was expressed in 5 out of the 20 interviews, with 3 of these comprising interviews with consultants.

> I think more and more, probably green travel plans by way of a condition on a permit will probably go by the way side, by virtue of them being challenged [C5].

6.3.8 Synthesis of Interview Findings

Using the interview responses, an assessment was made of each participant's level of support and confidence in the ability for travel plans at new residential developments to be successfully implemented and achieve their desired outcomes. For example, an interview participant who was considered to be both highly supportive and highly confident would have cited a range of benefits associated with travel plans for new residential developments, but little or no disadvantages. In addition, they would have seen limited challenges associated with implementation and expressed a view that travel plans are likely to increase in focus into the future. Given that this assessment of interview participants was somewhat subjective, the results can only be considered as indicative.

Figure 6.5 shows the levels of support and confidence among interview participants based on the assessment undertaken. Participants generally felt supportive of the concept, but were not necessarily all that confident in it. No clear pattern emerged by organisation type, although local government representatives did appear to be more confident in travel plans for new residential developments than other industry representatives, including those working in state government. Almost half of all interview participants (13 out of 30) seemed both supportive and confident in travel plans for new residential developments, compared with only eight participants who appeared supportive but neither confident nor unconfident, and six who seemed supportive but unconfident. Only two interview participants (one property developer and one consultant) appeared to be both unsupportive and unconfident in travel plans for new residential developments.

A summary of the interview findings by key topic is provided in Table 6.5.

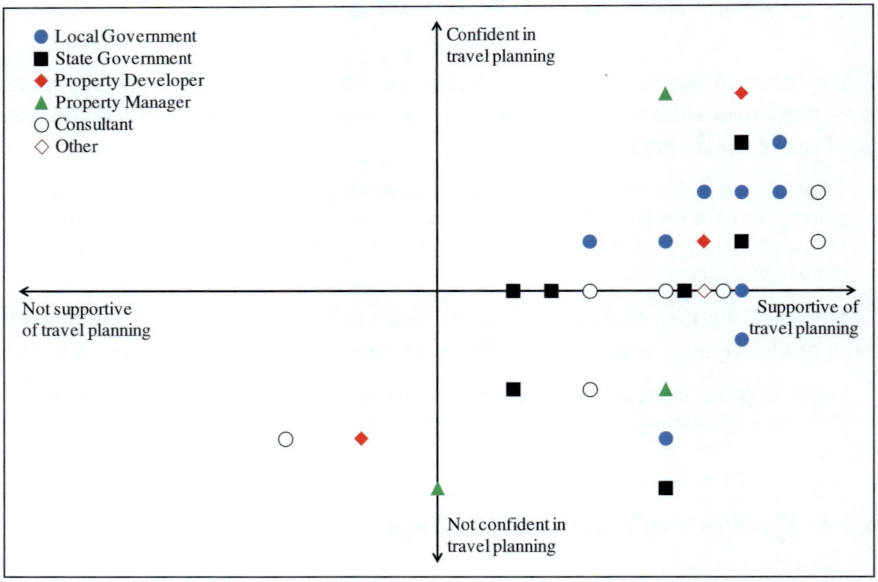

Fig. 6.5 Levels of support and confidence in travel plans for new residential developments. *Note* interview participants were not asked to indicate where they were positioned on the graph; an assessment was made by the researcher based on responses from interview participants to questions 3, 4, 7, 8 and 9

Table 6.5 Summary of interview findings

Interview participants had diverse backgrounds, yet most worked in planning or engineering roles
• Five out of 30 participants had no prior experience with travel plans, although 13 participants indicated a high level of travel planning experience (five or more years) • Property managers generally had little or no experience with travel plans
The industry has had little involvement with implementation and monitoring to date
• By their nature as authorities, local and state government were generally only involved in requiring travel plans and had limited involvement thereafter • Involvement was mostly focused on preparing/developing travel plans, rather than implementing or monitoring them
Travel plans for new residential developments were seen to offer a wide range of benefits
• Most commonly cited benefits were environmental and social benefits, less space required for car parking, and the process forcing developers to consider sustainable transport issues
Many potential shortcomings of travel plans for new residential developments were identified
• Most commonly cited shortcomings were a lack of follow-up or enforcement, lack of adequate transport infrastructure to support travel plans, lack of any government policy/legislative requirement in Australia, and a varying level of travel plan quality • In five of the interviews, participants felt there were no disadvantages associated with travel plans for new residential developments, as long as they are delivered effectively
Involvement of actors is focused on a small number of organisation types
• Consultants, local government, property developers and state government are the key actors currently involved in travel planning for new residential developments • Other actors involved in the process, but not interviewed, include architects, building managers, resident associations/committees, service providers and advocacy groups
Implementation was recognised by interview participants as a key challenge
• Most commonly cited challenges included the lack of any enforcement, uncertainty about implementation responsibilities, and a lack of ownership
Interview participants offered a diverse range of potential solutions to implementation challenges
• Most commonly cited solutions included developing a more robust planning/legal requirement, encouraging more resident engagement and ownership, involving the owners corporation in implementation, and having the developer fund the implementation of the travel plan • Views expressed that any planning legal/requirement should not be too prescriptive
Travel planning for new residential developments is expected to continue into the future
• Most participants felt that travel plans for new residential developments were either here to stay or will increase in focus, although a small proportion felt that they would reduce in scale

6.4 Discussion

This section compares the interview findings to the literature described earlier in Chap. 2 and discusses the implications for future travel planning practice. Where relevant, comparisons are also made to the findings from the survey of Victorian councils, reported in the previous chapter.

Interviews undertaken by Yeates and Enoch (2012) with developers in the United Kingdom highlighted cost savings as a key benefit of travel plans at new developments, yet only three out of the 20 interviews undertaken in this research (one each with local government, a property developer, and a property manager) identified this as a benefit. The reason for this difference is unclear but may reflect a greater level of experience and awareness in the United Kingdom of the cost savings offered by travel plans.

When asked about the potential disadvantages of travel plans at new residential developments, the most common responses related to a lack of follow-up or enforcement, lack of adequate transport infrastructure to support travel plans, lack of any government policy/legislative requirement, and a varying level of quality in travel plans that are prepared.

The issue of insufficient enforcement is raised by previous research (Enoch and Ison 2008; Llewellyn et al. 2014) noting that this can reduce the effectiveness of the travel plans introduced. It was also raised in the survey of Victorian councils (reported in Chap. 5) in which a lack of resources was identified as a key barrier to enforcing travel plans. The enforcement of travel plans for new residential developments is discussed further in Chap. 9.

Related to the issues concerning enforcement is that only two out of the 20 organisations that were interviewed indicated any involvement with monitoring travel plans for new residential developments. This finding is also consistent with the survey results reported in Chap. 5.

The lack of any government policy or legislative requirement for travel plans in Australia is in contrast to the situation in the United Kingdom where national planning policy gives specific reference to requiring travel plans for new developments (Rye et al. 2011a). Similar policies at a national level also exist in Sweden and Switzerland (Rye et al. 2011b). As detailed in Chap. 2, supportive planning policy is recognised as a key success factor in requiring travel plans for new developments (Addison & Associates 2008), with this need also identified in Chap. 5.

The issue of varying travel plan quality is consistent with research interviews undertaken by Enoch and Ison (2008, p. 24) in the United Kingdom who found that '…the increased number (and variable quality) of consultants adopting a standardised "*sausage machine*" approach to travel planning were seen by the interviewees as being especially problematic.' Poorer quality travel plans are generally expected to lead to less effective implementation, with this issue discussed further in the next chapter.

While not as pronounced, a number of other issues raised in the interviews were also consistent with the literature. These issues included lip service being paid to travel plans by developers solely for the purpose of seeking planning approval (Davison et al. 2010; Rye et al. 2011a), a lack of guidelines concerning travel plans for new developments in general (Wynne 2013), the resource intensive nature of travel plans when secured through the planning process (Davison et al. 2010; Wynne 2013), a lack of travel plan ownership (Yeates and Enoch 2012), and the travel plan requirement coming too late in the planning application process (Hendricks 2008). In commenting on the issue of lip service, Davison et al. (2010, p. 20) argue that '… where the motivation is limited to regulation, completing a travel plan can become a 'tick box' activity and create paperwork rather than effective action.'

While various implementation issues were identified, a number of opportunities for enhancing implementation were also proposed by interview participants. Examples included the development of an appropriate planning requirement, encouraging resident engagement and ownership in the process, and ensuring the developer funds the implementation of the travel plan. These, along with other opportunities, are discussed further in Chap. 9.

Finally, the view that travel plans for new residential developments are either here to stay or will increase in focus was expressed in most of the interviews, consistent with the literature (Davison et al. 2010; Llewellyn et al. 2014) and the survey findings from Chap. 5. This demonstrates that residential travel plans are valued and are likely to continue to be required for new developments.

6.5 Conclusion

The aim of this chapter is to gain an appreciation for the perspectives of industry actors involved in travel planning for new residential developments. In doing so, a series of interviews provided insight on their perceived benefits and shortcomings, extent of involvement and stakeholder interactions, implementation challenges and potential solutions, and future expectations.

The interview findings showed general support among industry representatives for travel plans at new residential developments, but limited confidence in the ability to implement them successfully. A number of challenges were identified with implementation, not least of which were the lack of enforcement, uncertainty over implementation responsibilities, and a general lack of ownership. A number of opportunities to address these challenges were identified by interview participants. These opportunities, along with others, are explored further in Chap. 9 where the theories of implementation and planning enforcement (introduced in Chap. 3) are applied to the research findings and developed further into an integrated theory.

The issue of varying travel plan quality was also raised by a number of interview participants. A particular focus is placed on this issue in the next chapter through an assessment of the quality of travel plans prepared for new residential developments.

References

Addison & Associates. (2008). *Delivering travel plans through the planning process—Research report*. London, UK: Department for Transport and Communities and Local Government.

Bryman, A. (2001). *Social research methods*. Oxford, UK: Oxford University Press.

Davison, L., Enoch, M., & Ison, S. (2010). *European experience of travel plans: An expert perspective*. Paper presented to 12th World Conference in Transportation Research, Lisbon, Portugal, July 2010.

De Gruyter, C., Rose, G., & Currie G (2015). 'Enhancing the impact of travel plans for new residential developments: Insights from implementation theory', *Transport Policy, 40,* 24–35.

Enoch, M., & Ison, S. (2008). *Expert perspectives on the past, present and future of travel plans in the UK: Research report*. Department for Transport and the National Business Travel Network.

Hendricks, S. (2008). Four challenges to incorporating transportation demand management into the land development process. *Transportation Research Record: Journal of the Transportation Research Board, no., 2046*, 30–36.

Hennink, M., Hutter, I., & Bailey, A. (2011). *Qualitative research methods*. UK: SAGE Publications Ltd.

Llewellyn, R., Tricker, R., & Paton, D. (2014). Travel plans: A critical comparison of the application of land use planning processes in England and Scotland. *Transport*, no. iFirst, pp. 1–13.

Mack, N., Woodsong, C., MacQueen, K., Guest, G., & Namey, E. (2005). *Qualitative research methods: A data collector's field guide*. North Carolina, USA: Family Health International.

Richardson, A. J., Ampt, E. S., & Meyburg, A. H. (1995). *Survey methods for transport planning* (1st edn). Eucalytpus Press.

Rye, T., Green, C., Young, E., & Ison, S. (2011a). Using the land-use planning process to secure travel plans: An assessment of progress in England to date. *Journal of Transport Geography, 19*(2), 235–243.

Rye, T., Welsch, J., Plevnik, A., & de Tomassi, R. (2011b). First steps towards cross-national transfer in integrating mobility management and land use planning in the EU and Switzerland. *Transport Policy, 18*, 533–543.

Wynne, L. (2013). *Travel planning and the land-use planning system: Understanding the effectiveness of travel planning requirements in NSW Development Control Plans*. Unpublished Masters Thesis, University of New South Wales, Australia.

Yeates, S., & Enoch, M. (2012). *Travel plans from the developer perspective*. Paper presented to 91st Transportation Research Board (TRB) Annual Meeting, Washington DC.

Yin, R. (2009). *Case study research: Design and methods* (4th edn). California, US: SAGE Publications.

Chapter 7
Travel Plan Quality

7.1 Introduction

The previous chapter presented the second set of research results to provide an appreciation of the perspectives of industry actors involved in travel planning for new residential developments. In doing so, it provided insight on their perceived benefits and shortcomings, extent of industry involvement and stakeholder interactions, implementation challenges and potential solutions, and future expectations. One of the key issues identified was the varying level of quality in travel plans that are prepared for new developments, including residential sites.

This chapter, as positioned in Fig. 7.1, places a focus on travel plan quality by detailing the findings of the desktop assessment of travel plans, corresponding to research component 3. Table 7.1 details the research gap, opportunity and objective associated with this research component.

In line with research objective 3, the aim of this chapter is to evaluate the quality of travel plans prepared for new residential developments (effectiveness is covered in Chap. 8). This is achieved by assessing a sample of travel plans prepared for new residential developments in Victoria against a best practice framework. Assessing travel plan quality is particularly relevant for new developments given that those responsible for implementation may not have been involved in preparing the travel plan, therefore underlining the need for a travel plan that is both clear and comprehensive. Moreover, understanding the quality of travel plans prepared for new residential developments can assist in identifying opportunities for improvement, thereby increasing the likelihood that they will be implemented successfully and achieve their objectives.

This chapter commences with an overview of the research literature relevant to evaluating travel plan quality, which is used to inform the development of a

C. De Gruyter, *Travel Plans for New Residential Developments: Insights from Theory and Practice*, Springer Theses,
DOI 10.1007/978-981-10-2092-6_7

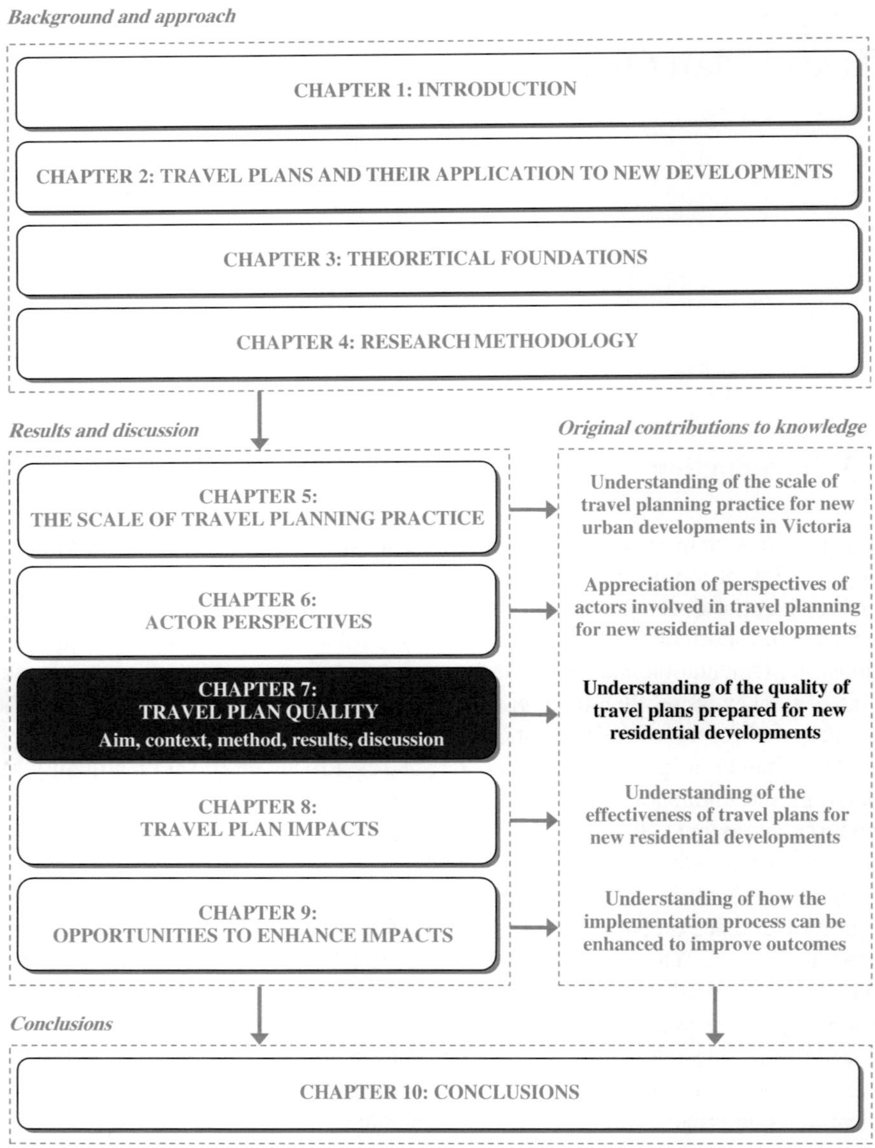

Fig. 7.1 Position of Chap. 7 in the thesis structure

framework for assessing the quality of the travel plans. The travel plans are then described in terms of their content, followed by a quantitative assessment of their quality. The chapter concludes by discussing the implications for future travel planning practice.

Table 7.1 Research gap, opportunity and objective associated with research component 3

Research gap →	Research opportunity →	Research objective →	Research component
No formal assessment of the quality of travel plans prepared for new residential developments has been undertaken	Undertake a quantitative assessment of the quality of travel plans prepared for new residential developments to help identify their relative merits and potential areas for improvement	3. To evaluate their quality and effectiveness	3. Desktop assessment of travel plans

7.2 Research Context

This section provides an overview of the literature relating to travel plan quality which is used to inform the development of an assessment framework in Sect. 7.3. However, given the relative paucity of literature available on travel plan quality that is specific to new developments, particularly residential sites, the review draws upon a wider set of literature covering travel plan quality for both new developments and pre-existing sites, regardless of land use type.

Firstly, and while not directly related to travel plans, Mansfield and Hartell (2012) provide a useful framework for assessing transport sustainability plans in the United States. They state that the tenets of plan quality include a vision statement, comprehensive fact base, consistent policy framework, clear implementation and monitoring procedures, accountability for the interdependence of actions, and open participation in the plan development process. Each of these tenets has relevance in the context of travel plans for new residential developments. Mansfield and Hartell (2012) also recognise that there is no 'one size fits all' approach to plan development, particularly when addressing complex and multi-faceted issues such as sustainability. This is again relevant to travel plans as it is recognised that their content needs to be tailored to the local context and transport needs of a site (Department for Transport 2009).

In the United Kingdom, Transport for London (2011a) developed an online tool called ATTrBuTE (Assessment Tool for Travel plan Reviewing, Building, Testing and Evaluation) to evaluate the quality of incoming travel plans as part of the planning process. The criteria included in the tool (presented as a series of questions) are designed to test the extent to which a travel plan has been prepared in accordance with their guidance on travel planning for new developments (Transport for London 2011b). A total of 11 categories are included in the tool, with a set of scored criteria under each category. The categories relate to background information about the development, references to policy, site assessments, objectives and targets, travel plan coordination, measures, monitoring, enforcement and funding. The travel plan needs to score above 70 % in order to pass the assessment.

In a separate set of guidelines on delivering travel plans through the planning process, the UK Department for Transport (2009) recognise that without a robust process for evaluating incoming travel plans, there is no basis to make a judgement as to whether the travel plan will meet its intended outcomes and will therefore be fit for purpose. It is also recognised that the methodology used to evaluate the quality of the travel plan should be made publicly available so those preparing travel plans are aware of the components that require particular attention.

Addison & Associates (2008) report on the experience of local authorities in the United Kingdom in evaluating travel plans submitted as part of the planning process. Common aspects that are examined when evaluating travel plan quality include:

- Evidence of a site assessment and baseline travel pattern information
- Inclusion of a comprehensive range of measures that are realistic and appropriate to the site
- Use of SMART targets (Specific, Measurable, Achievable, Realistic and Time-based)
- Commitment to implementation and monitoring with responsibilities clearly identified
- Nomination of a travel plan coordinator with contact details provided
- Clear procedures for monitoring and reviewing the travel plan on a regular basis.

WS Atkins (2002) designed a software tool for the UK Department for Transport to assess the process and content of workplace travel plans. The tool provides a scoring system with 14 categories that cover key aspects of travel plans. Similarly, the British Standards Institution (2008) provides a number of useful checks that can be used in evaluating the quality of a workplace travel plan for a new development. Aspects covered include aims and objectives, existing transport conditions, output and outcome targets, commitment to meeting the aims and objectives, and actions that are deliverable and funded.

In Australia, Wake et al. (2010) highlight the importance of evaluating the quality of travel plans against a good practice benchmark. They recommend checking a number of elements such as commitment to the travel plan, objectives and performance targets, baseline information, feasibility of actions, and the framework proposed for implementation and monitoring. Travel plan guidelines for new developments in the City of Darebin in Victoria state that it is necessary for the council to play a role at each stage in the evaluation of travel plans (PBAI 2005). A pro-forma is provided so that council staff can ensure that the travel plan meets various requirements at the planning application stage. Various councils in the United Kingdom, such as Wiltshire County Council have also developed similar pro-formas (Wiltshire County Council 2004).

Table 7.2 Synthesis of travel plan elements covered by the literature

Travel plan elements	Literature item (see key below table)										Total
	[A]	[B]	[C]	[D]	[E]	[F]	[G]	[H]	[I]	[J]	
Background information											
Type of land use/s		✓	✓	✓	✓	✓	✓	✓	✓	✓	9
Development address		✓	✓	✓		✓	✓	✓	✓	✓	8
Number and type of expected users		✓	✓	✓	✓	✓	✓	✓	✓		8
Contact details of travel plan author		✓	✓		✓	✓		✓	✓	✓	7
Development size		✓	✓	✓			✓	✓	✓	✓	7
Number of car parking spaces		✓	✓	✓	✓	✓		✓			6
Reference to relevant policies		✓	✓	✓	✓		✓	✓			6
Benefits of travel plan			✓	✓	✓	✓					4
Development name or site name			✓			✓	✓			✓	4
Development phasing		✓	✓	✓					✓		4
Number of bicycle parking spaces		✓	✓		✓			✓			4
Rationale for travel plan			✓	✓	✓		✓				4
Reference to relevant agreement and/or condition		✓	✓			✓		✓			4
Reference to relevant travel planning guidance		✓	✓		✓						3
Timescales for occupation			✓	✓				✓			3
Existing conditions											
Estimate of baseline travel patterns	✓	✓	✓	✓	✓	✓	✓	✓	✓	✓	10
Existing transport networks and services	✓	✓	✓	✓	✓	✓	✓	✓	✓	✓	10
Existing travel initiatives available	✓	✓		✓		✓	✓		✓	✓	7
Organisational policies and other initiatives			✓	✓		✓	✓				4
Amenities and facilities in surrounding area			✓		✓				✓		3
Objectives and targets											
Objectives reflective of site characteristics		✓	✓	✓	✓	✓	✓	✓	✓	✓	9
Targets linked to objectives	✓	✓	✓	✓	✓		✓	✓	✓	✓	9
Objectives reflective of relevant policy	✓	✓	✓	✓	✓		✓	✓	✓		8
SMART (specific, measurable, achievable, relevant, time-based)			✓		✓	✓	✓		✓	✓	6
Travel plan measures											
Alignment with objectives and targets		✓	✓	✓	✓	✓	✓	✓	✓		8
Description of measures			✓	✓	✓	✓	✓	✓	✓	✓	8
Consideration to all transport modes (including deliveries)			✓	✓	✓	✓	✓	✓	✓	✓	8
Reflective of characteristics and needs of site		✓	✓	✓	✓	✓	✓	✓	✓		8
Timescales	✓	✓	✓	✓	✓		✓		✓		7
Marketing and promotion			✓	✓	✓	✓			✓	✓	6

(continued)

Table 7.2 (continued)

Travel plan elements	Literature item (see key below table)										Total
	[A]	[B]	[C]	[D]	[E]	[F]	[G]	[H]	[I]	[J]	
Travel plan management											
Roles and responsibilities	✓	✓	✓	✓	✓		✓	✓	✓	✓	9
Commitment to implementation		✓	✓	✓	✓	✓	✓	✓	✓		8
Travel plan coordinator		✓	✓	✓	✓		✓	✓	✓	✓	8
Budget and funding stream for travel plan coordinator		✓	✓	✓	✓	✓	✓		✓		7
Budget and funding stream for travel plan measures		✓	✓	✓	✓	✓	✓		✓		7
Securing and enforcement	✓	✓	✓	✓				✓	✓	✓	7
Partnerships	✓		✓	✓	✓		✓			✓	6
Time allocated for travel plan coordinator		✓	✓	✓			✓	✓	✓		6
Handover arrangements (e.g. from developer to occupant)			✓	✓					✓	✓	4
Monitoring and review											
Frequency	✓	✓	✓	✓	✓	✓	✓	✓	✓	✓	10
Method	✓	✓	✓	✓	✓	✓	✓	✓	✓	✓	10
Timing	✓	✓	✓	✓	✓	✓	✓	✓	✓	✓	10
Roles and responsibilities	✓	✓	✓	✓	✓		✓	✓	✓	✓	9
Budget and funding stream		✓	✓	✓	✓	✓			✓	✓	7
Reporting format			✓	✓	✓	✓	✓			✓	6
Use of results				✓	✓	✓	✓		✓		5
Total elements	13	33	44	39	35	28	33	28	34	25	

Source Author's synthesis of the literature based on
[A] Mansfield and Hartell (2012)
[B] Transport for London (2011a)
[C] Transport for London (2011b)
[D] Department for Transport (2009)
[E] WS Atkins (2002)
[F] British Standards Institution (2008)
[G] Wake et al. (2010)
[H] PBAI (2005)
[I] Addison & Associates (2008)
[J] Wiltshire County Council (2004)

Table 7.2 provides a synthesis of the elements covered by the literature that are relevant to evaluating travel plan quality. The most comprehensive set of elements were covered by Transport for London (2011b) (column C of Table 7.2). Elements most commonly cited by the literature (by at least nine out of the ten items reviewed) related to:

- Background information on the type of land use/s
- Existing transport networks and services
- Baseline travel patterns
- Objectives that are reflective of the site's characteristics
- Targets linked to objectives

- Roles and responsibilities (both in terms of implementation and monitoring)
- Method, timing and frequency of monitoring and review.

Handover arrangements (e.g. from developer to occupant) were cited by only four items of literature, despite this being recognised as a key factor in ensuring effective implementation of the travel plan (Department for Transport 2009; Transport for London 2011b). However, contact details for the travel plan author, which help to facilitate handover arrangements, were referenced by seven of the ten literature items reviewed.

Finally, given the difficulty with estimating baseline travel patterns at new developments (since the occupier is usually unknown when preparing the travel plan), an assessment of the existing transport network, in conjunction with the use of secondary survey data (such as the census and other household travel survey data), are important for understanding existing conditions.

7.3 Research Method

This section describes the method that was used to assess the quality of a set of travel plans prepared for new residential developments in Victoria, Australia. This included the development of an assessment framework, sourcing copies of the travel plans and reviewing their content, and then applying the assessment framework to the travel plans.

7.3.1 Development of Assessment Framework

Taking into account the findings from the review of relevant literature (presented in Sect. 7.2), a framework was developed to assess the quality of travel plans for new residential developments. Consideration was given to including all relevant criteria while ensuring the framework could still be easily understood and applied. Specific characteristics of new developments were explicitly taken into account when developing the framework. For example, the framework asks whether an *estimate* of *expected* travel patterns has been made, rather than whether a baseline travel survey was undertaken, as the site's users are often unknown at the time of preparing the travel plan (Department for Transport 2009). It was also important to ensure that the framework did not contain any criteria that could be open to alternative or creative interpretation. The aim was to develop a framework that would result in consistent assessment outcomes when applied by others. Guidance was therefore provided in the framework on how scores should be assigned to each criterion.

The assessment framework is shown in Table 7.3. It contains six key headings (consistent with Table 7.2) which are expanded out to a total of 54 specific criteria. A scoring system is provided which incorporates implied weightings for each criterion to reflect their relative importance, as suggested by the research literature. The maximum total score available is 116 points.

Table 7.3 Assessment framework

Assessment criterion	Scoring
Background information	
1. Is relevant background information about the development included? (max 7 points)	
1.1 Is the address of the development provided?	No = 0, yes = 1
1.2 Are the types of land use/s stated (e.g. residential, education, commercial)?	No = 0, yes = 1
1.3 Is the size of the development stated (e.g. no. of residential dwellings)?	No = 0, yes = 1
1.4 Are the type/s of expected users stated (e.g. residents, students, employees)?	No = 0, yes = 1
1.5 Is the number of proposed car parking spaces stated?	No = 0, yes = 1
1.6 Is the number of proposed bicycle parking spaces stated?	No = 0, yes = 1
1.7 Are the expected date/s of occupation stated?	No = 0, yes = 1
2. Are relevant contact details provided? (max 5 points)	
2.1 Are contact details provided for the travel plan author? *(Organisation, address, contact name, phone number, email)*	No = 0, partially = 1, yes = 2
2.2 Are contact details provided for the development applicant? *(Organisation, address, contact name, phone number, email)*	No = 0, partially = 2, yes = 3
3. Is the rationale for the travel plan clearly stated? (max 6 points)	
3.1 Are reasons/motivations for the travel plan clearly stated?	No = 0, yes = 1
3.2 Is reference made to relevant policies and/or strategies?	No = 0, partially = 1, yes = 2
3.3 Is reference made to a relevant planning condition/agreement?	No = 0, partially = 2, yes = 3
Existing conditions	
4. Has a site audit been undertaken and appropriately documented? (max 9 points)	
4.1 Are the existing transport networks and services (all modes) reported?	No = 0, partially = 2, yes = 4
4.2 Are any existing organisational policies/initiatives specified (if applicable)?	No = 0, yes = 1, N/A = 1
4.3 Are transport issues and opportunities identified?	No = 0, partially = 2, yes = 4
5. Has an estimate of expected travel patterns been made? (max 8 points)	
5.1 Has an assessment been made of the likely travel behaviour of expected users?	No = 0, partially = 2, yes = 4
5.2 Is reference made to trip generation estimates?	No = 0, yes = 2
5.3 Are secondary data sources used (e.g. census data)?	No = 0, yes = 2
Objectives and targets	
6. Are a clear set of appropriate objectives identified? (max 6 points)	
6.1 Are the objectives linked to relevant policies and/or strategies?	No = 0, partially = 1, yes = 2
6.2 Are the objectives responsive to issues and opportunities facing the site?	No = 0, partially = 2, yes = 4

(continued)

Table 7.3 (continued)

Assessment criterion	Scoring
7. Are a clear set of appropriate targets identified? (max 8 points)	
7.1 Are targets focused on the outcomes of the travel plan (not process or outputs)?	No = 0, yes = 1
7.2 Are targets linked to the travel plan's objectives?	No = 0, partially = 1, yes = 2
7.3 Are targets informed by existing conditions?	No = 0, partially = 1, yes = 2
7.4 Do the targets contain SMART elements? (*Specific, measurable, achievable, relevant, time-based*)	None = 0, 1-3 elements = 1, 4-5 elements = 2
7.5 Are suitable accompanying indicators identified?	No = 0, yes = 1
Travel plan measures	
8. Is a package of suitable measures proposed? (max 13 points)	
8.1 Are the measures aligned with the objectives and targets identified?	No = 0, partially = 2, yes = 3
8.2 Is consideration given to all relevant modes (including trip substitution)?	No = 0, partially = 3, yes = 5
8.3 Are the measures likely to address the transport issues at the site?	No = 0, partially = 3, yes = 5
9. Is sufficient information provided to guide the implementation of each measure? (max 8 points)	
9.1 Is a description of each measure given?	No = 0, partially = 1, yes = 2
9.2 Is a timeframe for implementing each measure stated?	No = 0, partially = 1, yes = 2
9.3 Is the responsibility for implementing each measure stated?	No = 0, partially = 1, yes = 2
9.4 Is the cost of each measure specified?	No = 0, partially = 1, yes = 2
Travel plan management	
10. Is a clear statement of commitment provided? (max 8 points)	
10.1 Is commitment provided towards implementing the travel plan?	No = 0, partially = 2, yes = 4
10.2 Is commitment provided towards monitoring and reviewing the travel plan?	No = 0, partially = 2, yes = 4
11. Has a person been identified to manage/lead the travel plan (e.g. travel plan coordinator)? (max 8 points)	
11.1 Are contact details for a coordinator provided? (*Organisation, address, contact name, phone number, email*)	No = 0, partially = 2, yes = 4
11.2 Is the role and responsibilities of the coordinator clearly stated?	No = 0, partially = 1, yes = 2
11.3 Is an estimation of time allocated to the coordinator role clearly stated?	No = 0, partially = 1, yes = 2
12. Are the roles and responsibilities of any others clearly defined? (max 5 points)	
12.1 Is a working/steering group identified?	No = 0, yes = 1
12.2 Are partnerships with other stakeholders identified?	No = 0, yes = 1
12.3 Are handover arrangements (e.g. applicant to occupant) clearly stated?	No = 0, partially = 1, yes = 3

(continued)

Table 7.3 (continued)

Assessment criterion	Scoring
13. Is a sufficient budget included with funding streams identified? (max 8 points)	
13.1 Is a sufficient budget associated with the travel plan coordinator post specified?	No = 0, partially = 1, yes = 2
13.2 Is a sufficient budget associated with the travel plan measures specified?	No = 0, partially = 1, yes = 2
13.3 Is a sufficient budget associated with monitoring and review specified?	No = 0, partially = 1, yes = 2
13.4 Is justification given for the allocated budget?	No = 0, partially = 1, yes = 2
14. Is a plan for communications included? (max 3 points)	
14.1 Are communication updates with the site's users proposed?	No = 0, yes = 1
14.2 Is the use of branding/slogans proposed?	No = 0, yes = 1
14.3 Are events proposed to raise the profile of the travel plan (e.g. launch event)?	No = 0, yes = 1
Monitoring and review	
15. Is a clear process for monitoring and reviewing the travel plan included? (max 14 points)	
15.1 Is the timing and frequency (e.g. annual) of monitoring and review specified?	No = 0, partially = 1, yes = 2
15.2 Are responsibilities for undertaking monitoring and review stated?	No = 0, partially = 1, yes = 2
15.3 Is the cost associated with monitoring and review specified?	No = 0, partially = 1, yes = 2
15.4 Is the method of data collection specified (e.g. survey, counts)?	No = 0, partially = 1, yes = 2
15.5 Is the type of information to be sought specified (e.g. transport mode shares)?	No = 0, partially = 1, yes = 2

Figure 7.2 shows how the points are allocated across each of the key headings included in the assessment framework. In order to reflect the importance of the process through which the travel plan is managed and delivered, as well as the actual measures proposed in the travel plans, a greater percentage of points are allocated to these components (28 and 18 % respectively) compared with other key areas in the assessment framework (12–15 % each).

7.3.2 Sourcing of Travel Plans

Copies of travel plans prepared for new residential developments in Victoria in the last five years were initially sourced from councils. As the travel plan documents are normally in the public domain during the planning application process,

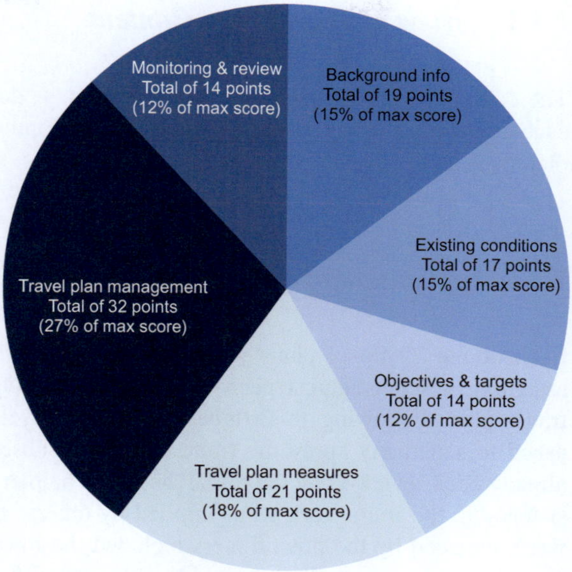

Fig. 7.2 Allocation of points by key headings used in the assessment framework

no confidentiality concerns were raised when sourcing the travel plans. However, due to representatives in some councils expressing difficulty with searching for travel plans within their internal databases, various consultants (authors of the travel plans) and the Victorian Department of Transport were also contacted with a request for travel plans prepared in the last 5 years. Overall, a total of 29 travel plans prepared specifically for new residential developments were sourced, yet this required considerable effort involving multiple reminders for assistance.

Travel plans prepared only in last 5 years were were requested to facilitate cooperation and reduce the level of burden for representatives from councils, consultancies and the Victorian Department of Transport. Regardless of this, most travel plans that were sourced had been prepared in later years, suggested limited activity more than 5 years ago. For example, 48 % of the travel plans sourced had been prepared in 2011–12, compared to 29 % in 2009–10, and only 16 % in 2007–08 (the remaining 7 % of travel plans were not dated).

Chapter 5 showed that around 100 travel plans had been required in Victoria in the 2-year period between 2010 and 2012. Assuming that this represented approximately half of the travel plans required in the last 5 years (consistent with the finding that 48 % of those sourced had been prepared during 2011–12), it can be estimated that around 200 travel plans have been required in the last 5 years. Therefore, the 29 travel plans used in the quality assessment would represent about 15 % of all travel plans in the last 5 years. As the council survey (reported in Chap. 5) did not request information about the number of travel plans required by land use type, it is not possible to estimate the true proportion of travel plans used in the quality assessment that were specific to new residential developments. However, a total of 29 travel plans was considered an adequate sample for assessment purposes.

7.3.3 Review of Travel Plan Content

The content of each travel plan was reviewed. A database was created to record each travel plan's key attributes such as development characteristics and travel plan measures.

7.3.4 Application of the Assessment Framework

Each of the 29 travel plans were analysed using the assessment framework by assigning a score against each criterion. This process took around 30 min per travel plan. Following this, three Australian travel planning practitioners were asked to separately apply the framework to a subset of the travel plans that had already been reviewed by the researcher. This helped to determine the level of consistency in the framework when applied by others. The subset of travel plans that were reviewed by the practitioners included the lowest, highest and average scoring travel plan, as initially assessed by the researcher.

7.4 Results

The results are presented in two main parts. The first part provides a summary of the content of the travel plans while the second part presents the results of the quality assessment.

7.4.1 Content Summary

7.4.1.1 Authorship and Document Length

Consultants prepared all of the travel plans, with the exception of a housing provider/manager who prepared one travel plan. The predominant service/discipline of the travel plan authors is shown in Table 7.4. Traffic engineering consultancies prepared more than half (59 %) of the travel plans. The average length of the travel plan documents was 11 pages, although this ranged from 1 to 38 pages.

7.4.1.2 Land Use Type and Location

All of the travel plans sourced were for new residential apartment buildings. However, around two-thirds (69 %) included other land uses as part of their development

Table 7.4 Travel plan authorship

Predominant service/discipline of organisational plan author	Number of travel plans	% of travel plans
Traffic engineering	17	59
Town planning	4	14
Architecture	3	10
Transport planning	2	7
Environmentally Sustainable Design (ESD)	2	7
Housing provision/management	1	3
Total	29	100

Table 7.5 Travel plans by development location (based on sample used in assessment)

Development location	Number of travel plans	% of travel plans
Inner metropolitan Melbourne	17	59
Middle metropolitan Melbourne	9	31
Outer metropolitan Melbourne	2	7
Regional Victoria	1	3
Total	29	100

application and were therefore classified as 'mixed-use'. In all of these cases though, the other land uses comprised only a very minor part of the overall development. Therefore, each of the developments were still predominantly residential. A common example was a development with retail at ground floor but several storeys of residential apartments above.

The remaining (31 %) travel plans that were sourced were prepared solely for new residential developments and therefore did not include any other land uses.

Table 7.5 shows that developments located in inner metropolitan Melbourne accounted for more than half (59 %) of the travel plans that were sourced. The middle metropolitan areas of Melbourne accounted for a smaller proportion (31 %) with some minor representation from outer metropolitan Melbourne (7 %) and regional Victoria (3 %). The spatial distribution of the developments is shown in Fig. 7.3. Most developments (around 85 %) were located within a 10 km radius of the Melbourne Central Business District (CBD), with good access to public transport, walking and cycling networks.

Chapter 5 showed that approximately 80 % of inner and middle metropolitan councils in Melbourne have previously required a travel plan for a new development, compared with only 20 % from outer metropolitan Melbourne and regional Victoria. This finding generally corresponds to the 29 travel plans that were sourced, suggesting they are broadly representative of Victoria in terms of their development location.

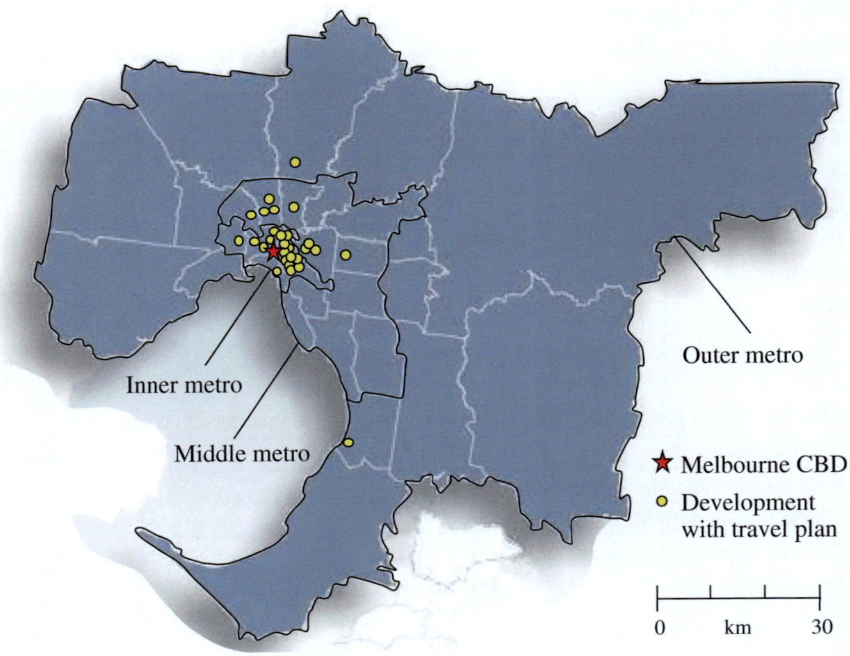

Fig. 7.3 Developments in metropolitan Melbourne with travel plans (based on sample)

7.4.1.3 Travel Plan Measures

Table 7.6 indicates the type and number of measures contained in the travel plans. On average, each travel plan contained around 11 measures, with information and infrastructure based initiatives making up more than half (57 %) of the total measures.

Table 7.7 provides a full list of actual measures contained in the travel plans. New resident kits, containing local information on sustainable transport options, were the most common measure proposed (by 93 % of travel plans). Bicycle parking was also relatively common (included in 90 % of travel plans). However,

Table 7.6 Types of measures included in the sample of travel plans

Type of measure	Average number of measures per travel plan	% of measures in travel plans
Information	3.5	33
Infrastructure	2.6	24
Incentive	1.8	17
Program	1.7	16
Other	1.1	10
Total	10.7	100

Table 7.7 Full list of individual measures included in the sample of travel plans

Travel plan measure	Number of travel plans that include measure	% of travel plans that include measure
Programs		
Events (e.g. Ride to Work Day)	12	41
Bicycle User Group (BUG)	11	38
Carpooling program	8	28
Cycling computer-based program	6	21
Pedometer-based walking program	6	21
Bike buddy scheme (pairing new riders with experienced riders)	3	10
Walking school bus	2	7
Personalised journey planning	1	3
Information		
New residents kit	27	93
Maps	23	79
Noticeboard/information display	20	69
Online information	14	48
Public transport timetables	13	45
New staff induction kit	3	10
Newsletters	2	7
Infrastructure		
Bicycle parking	26	90
Car sharing service	11	38
Directional signage	9	31
Showers	7	24
Change facilities	7	24
Car parking management (e.g. priority parking)	5	17
Lockers	4	14
Bicycle fleet	4	14
Hybrid cars	1	3
Incentives		
Free or discounted public transport tickets	24	83
Discounts from local shops (e.g. bicycle retailer)	8	28
Umbrellas	6	21
Car sharing membership	4	14
Free or subsidised bicycle	4	14
Cycling insurance cover	2	7

(continued)

Table 7.7 (continued)

Travel plan measure	Number of travel plans that include measure	% of travel plans that include measure
Free walker/cyclist breakfasts	1	3
Guaranteed ride home	1	3
Sustainable transport allowances	2	7
Parking cash-out	1	3
Other		
Investigating/lobbying for infrastructure improvements	10	34
Bicycle repair toolkit	6	21
Marketing/communications	8	28
Teleconferencing facilities	2	7
Videoconferencing facilities	1	3
Bicycle couriers for local deliveries	1	3
On-site bicycle maintenance service	1	3

bicycle parking is already a requirement under the Victorian Planning Provisions for residential developments of four storeys or more (Department of Transport, Planning and Local Infrastructure 2012). Other common travel plan measures included free or discounted public transport tickets (83 % of travel plans) and maps (79 % of travel plans).

It is also noted that some non-residential travel plan measures were included due to the mixed-use nature of most developments. Examples of such measures included new staff induction kits, lockers and teleconferencing facilities. These measures were generally associated with retail and office uses that were typically located on the ground floor of the developments.

7.4.2 Assessment Results

7.4.2.1 Overview

A summary of the results, in terms of the lowest, highest and average scoring travel plan (out of the 29 travel plans) is provided in Table 7.8. Across all travel plans, 47 % of the maximum possible score is achieved on average. The lowest scoring travel plan achieved only 22 % of the maximum possible score, with deficiencies relating to objectives and targets, travel plan management processes, and monitoring and review mechanisms. While the highest scoring travel plan addressed most of these deficiencies, there was still scope for improving the process for managing

Table 7.8 Summary of the assessment results

Assessment criterion	% of maximum possible score					
	Lowest scoring travel plan		Highest scoring travel plan		Average across all travel plans	
Background information						
1. Is relevant background information about the development included?	86	●	86	●	72	◕
2. Are relevant contact details provided?	80	●	80	●	66	◕
3. Is the rationale for the travel plan clearly stated?	0	○	50	◑	47	◑
Sub-total	56	◑	72	◕	62	◕
Existing conditions						
4. Has a site audit been undertaken and appropriately documented?	56	◑	100	●	67	◕
5. Has an estimate of expected travel patterns been made?	0	○	75	◕	33	◔
Sub-total	29	◔	88	●	51	◑
Objectives and targets						
6. Are a clear set of appropriate objectives identified?	0	○	100	●	56	◑
7. Are a clear set of appropriate targets identified?	0	○	88	●	51	◑
Sub-total	0	○	93	●	53	◑
Travel plan measures						
8. Is a package of suitable measures proposed?	46	◑	100	●	74	◕
9. Is sufficient information provided to guide the implementation of each measure?	38	◔	75	◕	68	◕
Sub-total	43	◑	90	●	71	◕
Travel plan management						
10. Is a clear statement of commitment provided?	0	○	38	◔	13	○
11. Has a person been identified to manage/lead the travel plan?	13	○	25	◔	19	○
12. Are the roles and responsibilities of any others clearly defined?	0	○	60	◑	28	◔
13. Is a sufficient budget included with funding streams identified?	0	○	0	○	14	○
14. Is a plan for communications included?	0	○	67	◕	31	◔
Sub-total	3	○	31	◔	19	○
Monitoring and review						
15. Is a clear process for monitoring and reviewing the travel plan included?	0	○	71	◕	45	◑
Sub-total	0	○	71	◕	45	◑
TOTAL	22	◔	69	◕	47	◑

○ = 0–20 %; ◔ = 21–40 %; ◑ = 41–60 %; ◕ = 61–80 %; ● = 81–100 %

the delivery of the travel plan. Furthermore, the highest scoring travel plan achieved only 69 % of the maximum possible score. Therefore, if applying the 70 % 'pass' criterion used in Transport for London's ATTrBuTE tool (Transport for London 2011a), none of the Victorian travel plans that were assessed would receive a 'pass'.

7.4.2.2 Key Strengths

Key strengths of the travel plans included:

- **Provision of background information**: this was generally documented appropriately, with relevant contact details provided for follow up purposes.
- **Detail provided about the site audit**: existing transport networks and services, plus key transport issues and opportunities, were sufficiently detailed and relevant to each site.
- **Appropriateness of travel plan measures**: measures were mostly tailored to the needs of each site, with sufficient information provided to guide their implementation.

7.4.2.3 Key Areas for Improvement

Key areas identified to improve the quality of the travel plans include:

- **Estimating expected travel patterns**: while future users of a proposed development are often unknown at the time of preparing a travel plan, this does not prohibit an indicative assessment to be made based on the existing transport network and services and the use of census data (which includes journey to work and car ownership data) or Victorian household travel survey data, both of which are freely available online (Australian Bureau of Statistics 2011; Department of Transport, Planning and Local Infrastructure 2010).
- **Specifying how the travel plan will be managed**: while the exact roles of individuals may be unknown at the time of preparing the travel plan, this does not preclude the specification of required roles. In addition, a commitment from the developer can be made, particularly in terms of funding towards implementing and monitoring the travel plan.
- **Outlining clear processes for monitoring and review**: clear details on how the travel plan will be monitored and reviewed are required to enable this to be undertaken effectively. This can include details relating to timing, frequency, responsibilities, cost and method. It can also specify what type of data will be collected, and how it will be reported and used.

7.4.2.4 Distribution of Scores

Figure 7.4 presents a cumulative frequency distribution of the scores achieved across the 29 travel plans that were reviewed. The figure shows that 15 of the 29

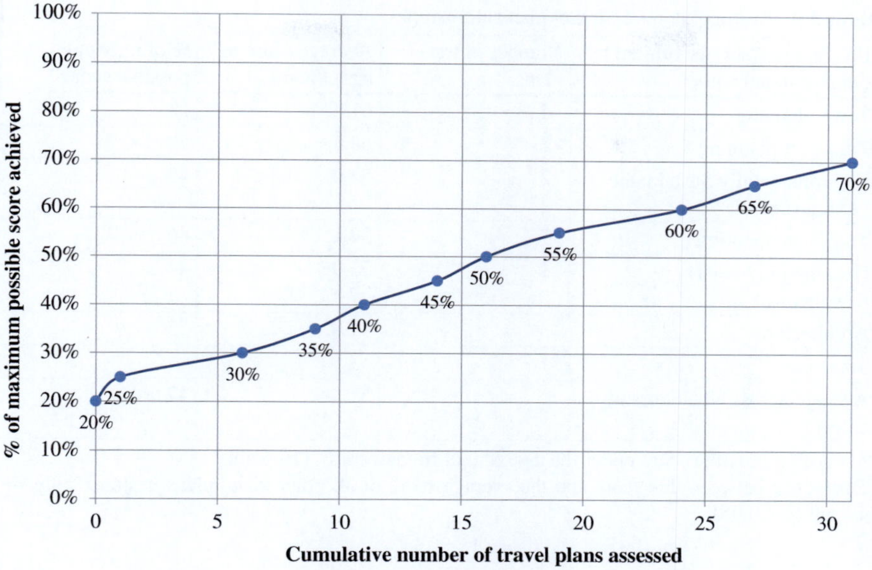

Fig. 7.4 Cumulative frequency distribution of maximum possible scores achieved

travel plans achieved 50 % or less of the maximum possible score. Overall, the distribution is relatively linear indicating a consistent spread of scores.

7.4.2.5 Scores by Travel Plan Authorship

It is also worth reflecting on whether travel plan quality differed according to the discipline of the travel plan author. Table 7.9 presents a summary of average scores by travel plan authorship showing that town planning consultancies produced the highest quality travel plans, while architectural firms produced the lowest quality travel plans. While these results are based on a relatively small sample, the difference in the average score of travel plans prepared by town planning consultancies and all other travel plan authors was statistically significant ($p < 0.05$). Furthermore, the difference in the average score of travel plans prepared by architectural firms and all other travel plan authors was also statistically significant ($p < 0.05$).

7.4.2.6 Scores by Travel Plan Document Length

The relationship between travel plan quality and document length was also assessed, as shown in Fig. 7.5. Longer travel plan documents were generally associated with higher scores and were therefore deemed to be of higher quality. Yet

Table 7.9 Average scores by travel plan authorship

Predominant service offered by organisational author	Number of travel plans	Average score of travel plans	% of maximum possible score[a]
Town planning	4	67.8[b]	58
Transport planning	2	68.5	59
Environmentally Sustainable Design (ESD)	2	53.0	45
Traffic engineering	17	54.1	46
Housing provision/ management	1	52.0	44
Architecture	3	39.0[b]	33
Total	31		
Average across all authorship categories		55.3	47

[a]Maximum possible score under the assessment framework is 116 points
[b]Difference between this score and the average score of all other travel plans is statistically significant ($p < 0.05$)

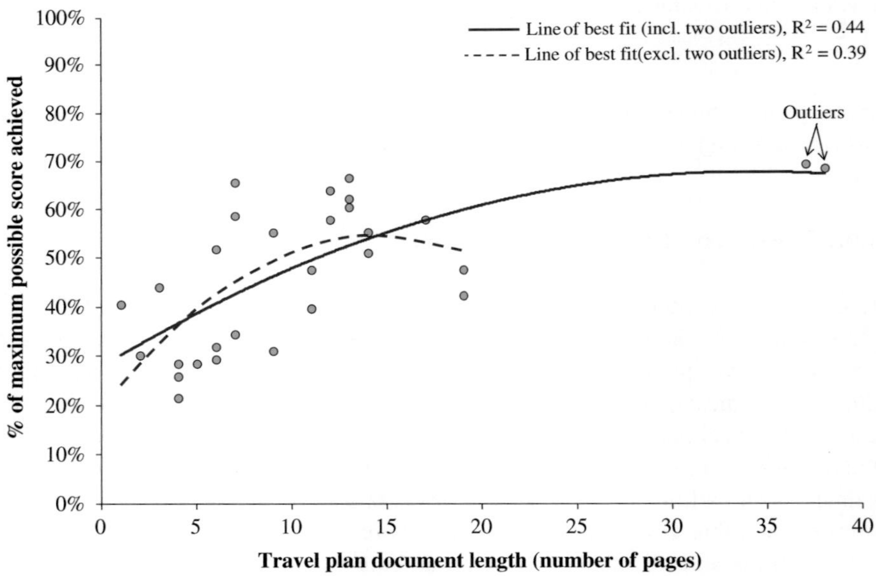

Fig. 7.5 Scores by travel plan document length

after removing two outliers from the analysis (travel plans that were considerably longer than others), quality declined slightly for travel plan documents of more than 15 pages. However, these findings should be interpreted with caution due the modest level of statistical correlation (R^2) achieved, ranging from only 0.39–0.44.

Table 7.10 Comparison of scores among travel planning practitioners

Travel plan reviewer	Score for 1st travel plan	Score for 2nd travel plan	Score for 3rd travel plan	Average score
Travel planning practitioner 1	17	36	86	46.3
Travel planning practitioner 2	31	63	74	56.0
Travel planning practitioner 3	23	46	61	43.3
Researcher	26	60	81	55.7
Average	24.3	51.3	75.5	50.3
Coefficient of variation (CV)	0.24	0.25	0.14	0.13

Intraclass Correlation Coefficient (ICC) = 0.86; level of variability considered acceptable if ICC > 0.80

7.4.2.7 Consistency in Applying the Assessment Framework

To check the level of consistency in applying the framework, scores from the three travel planning practitioners who applied the framework to a subset of the travel plans are presented in Table 7.10. There are two key observations to note from this table. Firstly, the ranking of travel plans (from lowest to highest scores) is consistent among each travel plan reviewer. Secondly, the variability in scores for each travel plan appears to be relatively low as scores do not depart substantially from the mean. An exception to this however would be the second travel plan where scores ranged from 36 to 63 (average of 51.3).

No clear pattern emerged in terms of specific criteria that were scored differently by reviewers for a given travel plan. However, scores for the second travel plan showed the largest differences for criteria relating to 'travel plan management', in particular whether a statement of commitment was provided in the travel plan (criteria 10.1 and 10.2 in Table 7.3).

To gain a better understanding of whether the level of variability in scores can be considered acceptable, a test for 'inter-rater reliability' was undertaken. Inter-rater reliability refers to the extent to which two or more raters (or in this case, the travel plan reviewers) agree when rating/scoring the same set of items (in this case, travel plans) (Wuensch 2007). The Intraclass Correlation Coefficient (ICC) provides a measure of inter-rater reliability and generally ranges from 0 to 1, with a score of more than 0.80 indicating an acceptable level of agreement between raters (Graham et al. 2012). For the scores presented in Table 7.10, an ICC of 0.86 was achieved. This provides confidence that the assessment framework can result in a sufficiently similar outcome when applied by different practitioners.

7.5 Discussion

The results of the quality assessment have shown that travel plans for new residential developments in Victoria are mostly being prepared by consultants, reflecting the interview findings reported in Chap. 6. While this may be appropriate due to the specific skills required, it may also impact upon the level of 'buy-in' and ownership of the travel plan by the developer. This is particularly relevant since ownership of a travel plan is deemed critical to its success (Howlett and Watson 2010). However, Victoria is certainly not alone in this situation, with Harrison (2003) providing an apt description of this issue in the context of the United Kingdom:

> …travel plans are increasingly being drafted for applicants by consultants. While this is welcome, in that a body of knowledge and expertise is being built up by specialists, it carries the risk that no one in the applicant's organisation has any particular personal commitment to making the plan a success. Indeed the individual who may feel most committed to the travel plan, having drafted and negotiated it, may be the consultant who will have no further connection with the site once planning permission has been granted (Harrison 2003, p. 400).

Furthermore, if a planning condition only requires the submission of a travel plan without any minimum quality standard, as is the case in Victoria, travel plans that score relatively low on the assessment may be approved in their current form as they would still technically meet the requirements of the planning condition. In these instances, there may not be a sufficient level of incentive for a developer to propose a more comprehensive travel plan.

The results of the assessment showed that only 47 % of the maximum possible score was achieved on average. Considerable scope therefore exists to improve the quality of travel plans prepared for new residential developments. For example, a greater focus could be placed on specifying how the travel plan will be managed and implemented beyond occupation of the development. In addition, proactive use of an assessment framework by councils to evaluate the quality of travel plans, with the process made transparent to property developers from the outset, would help to increase the likelihood that the travel plans will be implemented successfully and achieve their objectives. These, along with other opportunities for enhancement, are discussed further in Chap. 9.

It is also worth noting the limitations that exist in the assessment framework. Firstly, while informed by the research literature and adapted to local conditions, the scoring system was developed solely by the researcher. In reality, there may be some difference in opinion on the magnitude of scores set for particular criteria. Secondly, while the test for inter-rater reliability was considered acceptable, the wording of criteria relating to 'travel plan management' could be refined to enable a more consistent interpretation by travel planning practitioners. Thirdly, as recognised by Mansfield and Hartell (2012), the research method is limited to information contained in the travel plan documents that were reviewed and therefore does not capture undocumented but relevant practices. Despite these limitations, the application of the framework has highlighted a number of key areas for improving the quality of travel plans for new residential developments, which can ultimately contribute towards enhancing their effectiveness.

7.6 Conclusion

The aim of this chapter is to evaluate the quality of travel plans prepared for new residential developments. This was achieved by quantitatively assessing 29 travel plans prepared for new residential developments in Victoria against a best practice framework.

Results of the assessment showed that greater efforts need to be placed into estimating expected travel patterns of future users, specifying how the travel plan will be managed appropriately, and outlining clearer processes for monitoring and reviewing the travel plan. Using these and other findings from the research, Chap. 9 discusses a number of opportunities for enhancing the quality and subsequent effectiveness of travel plans for new residential developments.

The next chapter develops our understanding of the effectiveness of travel plans for new residential developments, based on a set of case study sites located in Melbourne.

References

Addison & Associates. (2008). *Delivering travel plans through the planning process—Research report*. London, UK: Department for Transport and Communities and Local Government.

Australian Bureau of Statistics. (2011). *Census of population and housing data*. Retrieved September 25, 2014, from http://www.abs.gov.au/census.

British Standards Institution. (2008). *PAS 500: 2008—National specification for workplace travel plans*. London, UK.

Department for Transport. (2009). *Good practice guidelines: Delivering travel plans through the planning process*. London, UK: Department for Transport.

Department of Transport, Planning and Local Infrastructure. (2010). *Victorian integrated survey of travel and activity data*. Retrieved September 25, 2014, from http://www.dtpli.vic.gov.au/.

Department of Transport, Planning and Local Infrastructure. (2012). *Victoria Planning Provisions (VPP)*. Retrieved September 25, 2014, from http://planningschemes.dpcd.vic.gov.au/schemes/vpps.

De Gruyter, C., Rose, G., & Currie G (2014). 'Methodology for Evaluating Quality of Travel Plans for New Developments', *Transportation Research Record: Journal of the Transportation Research Board,* No. 2417, 46–57.

Graham, M., Milanowski, A., & Miller, J. (2012). *Measuring and promoting inter-rater agreement of teacher and principal performance ratings*. US: Center for Education Compensation Reform (CECR).

Harrison, J. (2003). Travel plans and the planning system. *Journal of Planning and Environment Law*, April 2003, 397–403.

Howlett, R., & Watson, T. (2010). *Travel planning in Victoria—a new strategic approach to sustaining communities*. Paper presented to 33rd Australasian Transport Research Forum (ATRF), Canberra, Australia.

Mansfield, T. & Hartell, A. (2012). *Institutionalizing sustainability at the State DOT Level: A quantitative assessment of transportation sustainability plan quality*. Paper presented to Transportation Research Board (TRB) 2012 Annual Meeting, Washington DC.

PBAI. (2005). *Guidelines for the application and implementation of travel plans for new development in Darebin*. Victoria, Australia: City of Darebin.

Transport for London. (2011a). *ATTrBuTE v3 User Guide*, London, UK. Retrieved September 25, 2014, from http://www.attrbute.org.uk/.

Transport for London. (2011b). *Travel planning for new development in London—incorporating deliveries and servicing*. UK: Transport for London.

Wake, D., Thom, A. & Cummings, R. (2010). *Evaluating workplace travel plans*. Paper presented to 33rd Australasian Transport Research Forum (ATRF), Canberra, Australia.

Wiltshire County Council. (2004). *Development related travel plans in Wiltshire: Good practice guide*.

WS Atkins. (2002). *Workplace Travel Plan Evaluation Tool v2.7*. Department for Transport. Retrieved September 25, 2014, from http://www.imsaho.com/miscellaneous/travel_plan_evaluation_tool.asp.

Wuensch, K. (2007). *Inter-Rater Agreement*. North Carolina, US: Department of Psychology, East Carolina University,

Chapter 8
Travel Plan Impacts

8.1 Introduction

The previous chapter presented the third set of research results to provide an understanding of the quality of travel plans prepared for new residential developments. In doing so, it assessed a set of travel plans against a best practice framework developed in accordance with the literature.

This chapter, as positioned in Fig. 8.1, focuses on the effectiveness of travel plans by evaluating their impacts at a set of case study sites, corresponding to research component 4. Table 8.1 details the research gap, opportunity and objective associated with this research component.

In line with research objective 3, the aim of this chapter is to evaluate the effectiveness of travel plans for new residential developments (quality was covered in Chap. 7). Key aspects covered include:

- Vehicle trip generation rates
- Transport mode shares
- Car and bicycle parking utilisation
- Awareness and use of travel plan measures
- Residential self-selection effects.

This chapter begins by providing some additional research context through a literature review of residential self-selection. In the context of this research, residential self-selection may occur where residents choose to live or 'self-select' into a development with a travel plan because it is consistent with their existing attitudes and preferences towards more sustainable forms of transport. Therefore, any differences in travel behaviour that are observed when comparing to secondary data or control sites may be simply due to self-selection effects and not the travel plan itself. Following the literature review of residential self-selection, this chapter

C. De Gruyter, *Travel Plans for New Residential Developments: Insights from Theory and Practice*, Springer Theses,
DOI 10.1007/978-981-10-2092-6_8

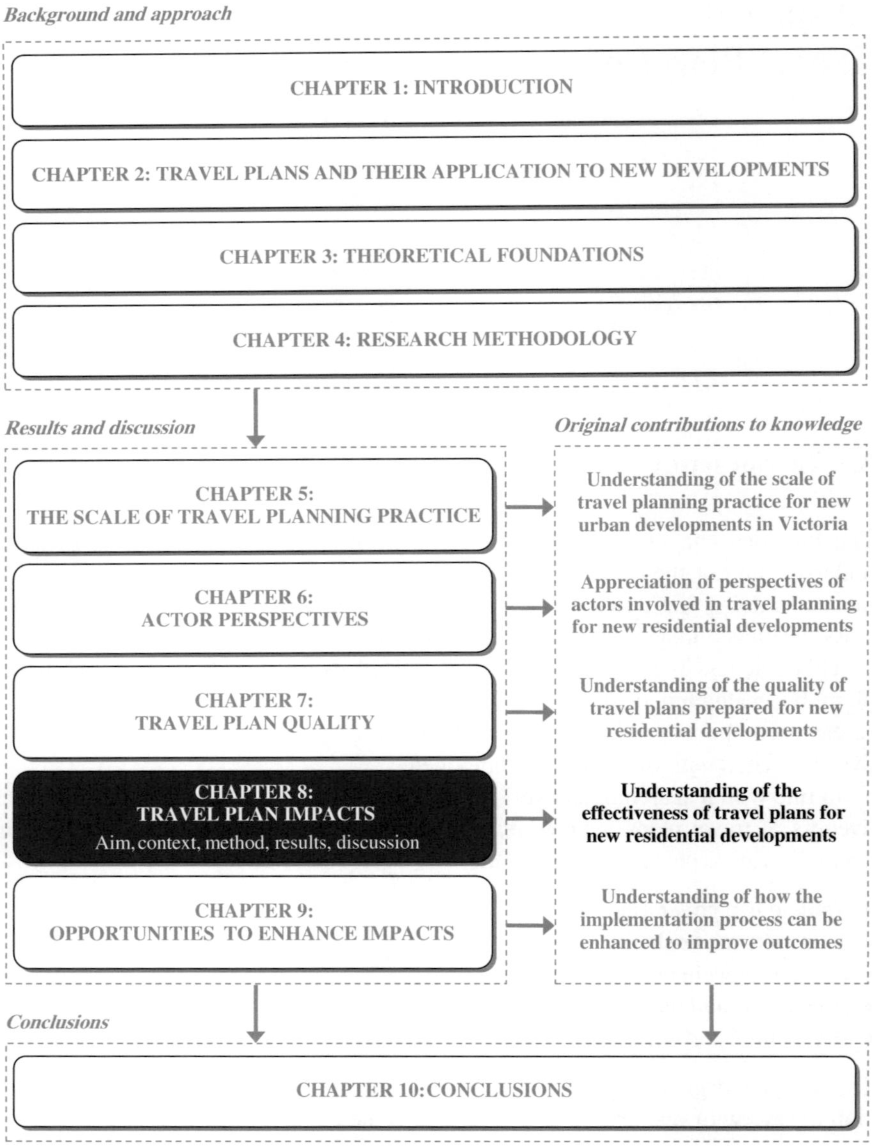

Fig. 8.1 Position of Chap. 8 in the thesis structure

describes the research methods used to evaluate the impacts of travel plans, including residential self-selection effects, at a set of case study sites. This is followed by the results, along with a discussion of their implications for future travel planning practice.

Table 8.1 Research gap, opportunity and objective associated with research component 4

Research gap →	Research opportunity →	Research objective →	Research component
Little research has been undertaken to appropriately quantify the effectiveness of travel plans in reducing car use at new residential developments, with no studies accounting for self-selection effects	Using a case study approach, evaluate the impacts of travel plans for new residential developments including self-selection effects to understand their effectiveness in reducing car use	3. To evaluate their quality and effectiveness	4. Case studies of new residential developments

8.2 Literature Review of Residential Self-selection

This section presents an overview of the literature relating to residential self-selection which provides context for assessing this phenomenon in the context of travel plans for new residential developments. The literature concerning other aspects relating to the evaluation of travel plan effectiveness was previously covered in Chap. 2 (see Sect. 2.6) so is not presented here.

While residential self-selection is yet to be investigated in the context of travel plans for new residential developments, research into carfree housing suggests that self-selection may contribute to a lower rate of observed car use (Broaddus 2010; Melia 2009; Melia et al. 2013). For example, Melia et al. (2013) found that potential demand for carfree housing in the United Kingdom is concentrated among 'carfree choosers' (those living without a car by choice), who are more likely to be younger, living in single person households and located in inner city areas. However, despite the potential for self-selection at carfree developments, the contribution that it makes towards lower rates of observed car use is yet to be quantified.

A detailed review undertaken by Mokhtarian and Cao (2008) discusses methods for assessing residential self-selection, ranging from direct questioning to structural equation modelling. This work was progressed by Cao et al. (2009) through a review of empirical findings from 38 studies that investigated residential self-selection in the context of the built environment. They found that while virtually every study reported a statistically significant influence of the built environment on travel behaviour after controlling for self-selection, the 'practical importance of that influence was seldom assessed' (Cao et al. 2009, p. 359). Where it was assessed, they found that residential self-selection accounted for 10–42 % of the variation in travel behaviour.

More recently, a technique known as Propensity Score Matching (PSM) has been applied to explicitly quantify the relative contributions that the built environment and self-selection have made on travel behaviour (Cao et al. 2010; Lee et al.

2014). PSM involves matching each observation in the treatment group with an almost identical observation in the control group, based on their propensity score (Caliendo and Kopeinig 2008). In the context of residential self-selection, the propensity score represents the probability of residing in the treatment group based on a set of attitudinal, preference and demographical characteristics (Cao et al. 2010; Lee et al. 2014; Mokhtarian and Cao 2008; Naess 2009). The average difference in travel behaviour between matched pairs then represents the effect of the treatment, after controlling for self-selection.

In applying PSM to regional travel survey data from North Carolina, Cao et al. (2010) found that the built environment generally plays a more important role in influencing driving behaviour than residential self-selection. When comparing driving behaviour between residents living in urban areas compared to those in other areas, self-selection was found to account for 15–24 % of vehicle miles driven per day.

In another study, Lee et al. (2014) used PSM to assess travel behaviour among urban and suburban baby boomers in Boston finding a very small self-selection effect (1–7 %) on automobile commuting, recreational non-motorised travel and utilitarian trips. However, for public transport commuting, they found a much larger contribution of self-selection of 43 %, suggesting the presence of a transit-oriented baby boomer market segment.

8.3 Research Methods

This section describes the methods used to evaluate the impacts of travel plans at a set of case study sites located in Melbourne, which includes an assessment of residential self-selection. A total of four case study sites with travel plans were adopted, termed as ‘case’ sites. For each case site, a matching ‘control’ site was chosen that had similar characteristics but no travel plan. While additional sites would help to provide a stronger understanding of travel plan effectiveness, only four case sites and four control sites could be evaluated within the resources available, particularly given that existing data on travel patterns was not available for any of the sites.

Specific data collection techniques adopted at each case and control site included:

- **Multi-modal person trip counts**: to provide information on vehicle trip generation rates and transport mode shares at each site.
- **Car and bicycle parking utilisation surveys**: to provide information on parking demand, supply and utilisation at each site.
- **Resident travel survey**: to understand levels of awareness and take-up of travel plan measures among residents, and to inform the assessment of residential self-selection.

Ethics approval was provided by the Monash University Human Research Ethics Committee (MUHREC) prior to collecting data at the sites (reference number CF12/1205—2012000586).

A description of the case and control sites is provided below. Further detail is then provided on the data collection undertaken at each site, along with the techniques used to analyse the data.

8.3.1 Description of Case and Control Sites

Only four new residential developments could be found in Melbourne that were built and occupied with travel plans that had been implemented, so all of these were selected as case sites. The selection of matching control sites involved a considerable number of site visits and discussions with property managers to ensure they were appropriate. While options for control sites will always be limited to what is available, best efforts were made to ensure sites were matched on their location, average dwelling size, on-site car parking provision, proportion of owner-occupiers, and the year that occupation commenced.

The location of each case and control site is shown in Fig. 8.2. Each control site was within at most 200 m of its corresponding case site, thereby providing a similar level of access to transport networks and services. All of the case and control sites were multi-storey apartment buildings, generally located within five kilometres of Melbourne's CBD, with good access to public transport, walking and cycling networks.

Key characteristics of the case and control sites are detailed in Table 8.2. The rate of on-site car parking was substantially lower at case sites 3 and 4 than their corresponding control sites. However, in these cases, reduced car parking was a feature of the travel plans. Similarly, the rate of on-site bicycle parking was not necessarily consistent between case and control sites as these facilities were typically provided as part of the travel plan. The proportion of residents that were owner-occupiers at each paired case and control site was relatively similar (with the exception of case-control 4) as was the average dwelling size and year that occupation commenced.

In addition, each site had similar car parking arrangements, in that on-site car parking spaces were allocated to individual dwellings, with the cost built into the purchase price of each dwelling. The only exception to this was case site 4 which did not provide any on-site car parking as part of its travel plan. As each control site was within at most 200 m of its corresponding case site, on-street parking restrictions were also similar between case and control sites.

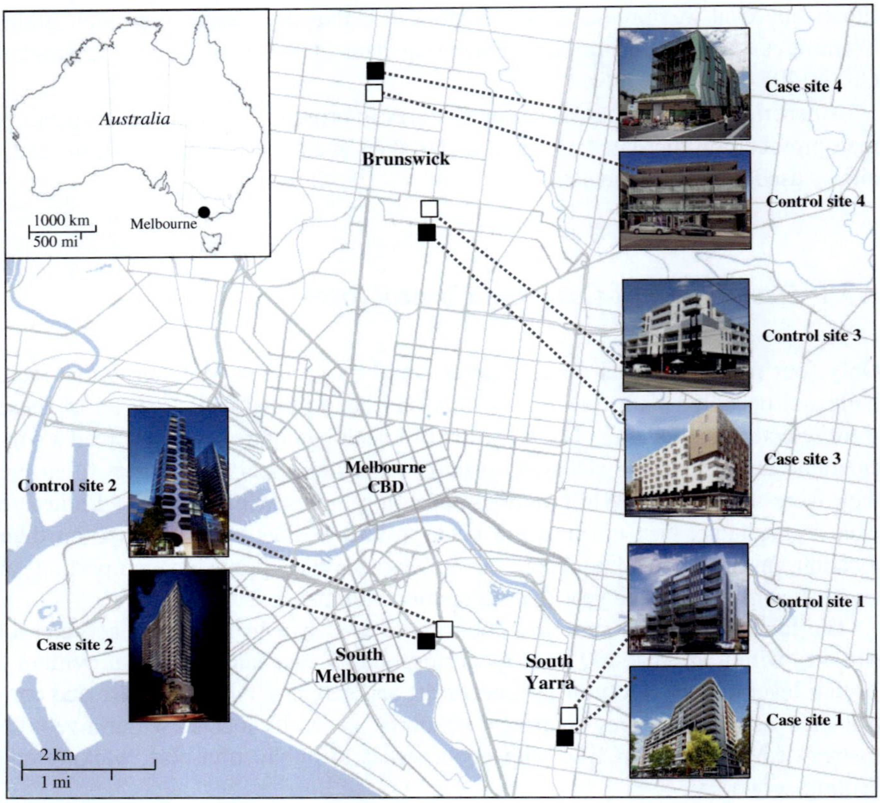

Fig. 8.2 Location of case and control sites in Melbourne, Australia

8.3.2 Data Collection and Analysis

8.3.2.1 Multi-modal person trip counts

Multi-modal person trip counts were conducted at the sites from March to May 2014, outside of school and public holidays. To ensure comparability, data for a given case site was collected at precisely the same time as its corresponding control site. The counts involved recording the number of people entering and leaving each access point of each site, by transport mode, on a Tuesday (7 am–9 am), Thursday (7 am–9 am) and Saturday (10 am–1 pm) within the same single week. People accessing the sites by car but parking on the street were recorded as far as observable.

The timing of the multi-modal counts corresponded to peak travel periods in Melbourne and captured both commute and non-commute trips (Department of Transport, Planning and Local Infrastructure 2010). However, these did not represent 'true' multi-modal counts in that they were conducted at each building access

Table 8.2 Key characteristics of the case and control sites

Site ID	Characteristic	Case site (travel plan)	Control site (no travel plan)	Key travel plan initiatives at case site
Case-Control 1 *Sites located approx 150 m apart*	Dwellings (average bedrooms/dwelling)	242 (1.4)	54 (1.5)	• Bicycle fleet (20 bicycles) • On-site car sharing vehicle with free membership • New resident kit containing local transport information • Free weekly public transport ticket • Transport information on building website • Online forum for organising carpooling • Umbrellas at lobby to encourage walking
	Car spaces (average car spaces/dwelling)	167 (0.7)	43 (0.8)	
	Bicycle spaces (average bicycle spaces/dwelling)	66 (0.3)	17 (0.3)	
	% owner-occupiers (%)	15	20	
	Year of occupation commencement	2010	2013	
Case-Control 2 *Sites located approx 40 m apart*	Dwellings (average bedrooms/dwelling)	282 (1.7)	156(1.9)	• 3 on-site car sharing vehicles with free membership • Transport information on building website • Free membership to Melbourne Bike Share
	Car spaces (average car spaces/dwelling)	268 (1.0)	158 (1.0)	
	Bicycle spaces (average bicycle spaces/dwelling)	80 (0.3)	129 (0.8)	
	% owner-occupiers	50	50	
	Year of occupation commencement	2013	2013	
Case-Control 3 *Sites located approx 190 m apart*	Dwellings (average bedrooms/dwelling)	124 (1.5)	45 (1.3)	• Additional on-site bicycle parking • Reduction in on-site car parking • New resident kit containing local transport information • Free weekly public transport ticket • Free membership to car sharing service • Transport information display in lobby
	Car spaces (average car spaces/dwelling)	97 (0.8)	57 (1.3)	
	Bicycle spaces (average bicycle spaces/dwelling)	110 (0.9)	18 (0.4)	
	% owner-occupiers (%)	30	35	
	Year of occupation commencement	2013	2012	

(continued)

Table 8.2 (continued)

Site ID	Characteristic	Case site (travel plan)	Control site (no travel plan)	Key travel plan initiatives at case site
Case-Control 4 *Sites located approx 200 m apart*	Dwellings (average bedrooms/dwelling)	24 (1.7)	34 (1.8)	• Additional on-site bicycle parking • No on-site car parking • Car sharing vehicle with free membership • 12 h free use of car sharing service per resident • After-ride shower facilities adjacent to bicycle parking • Building users guide containing transport information
	Car spaces (average car spaces/dwelling)	0 (0.0)	40 (1.2)	
	Bicycle spaces (average bicycle spaces/dwelling)	62 (2.6)	15 (0.4)	
	% owner-occupiers (%)	40	15	
	Year of occupation commencement	2013	2011	

point and therefore could not determine whether a person used public transport. Therefore, observed walking trips were assumed to incorporate any onward trips made by public transport.

Multiple surveyors were required to collect data at the sites (up to six at any given point in time). This was due to the presence of multiple access points that could not be observed from a single location, along with the need to collect data at a given case site at the same time as its corresponding control site. This resource requirement was supported by 16 undergraduate civil engineering students from Monash University, who assisted with data collection at the sites as part of a tailored assignment for one of their transport engineering units.

Table 8.3 details the total number of person trips, across all transport modes, observed at each case and control site during the survey periods. At least 100 person trips were recorded at each site in total, with close to 1,000 person trips observed at some of the larger sites. The number of person trips observed per dwelling was similar across all case sites (3.5 trips per dwelling) when compared to all control sites (3.4 trips per dwelling). However, person trip rates did vary between individually paired case and control sites. The reason for this is unknown but may be reflective of variations in residential occupancy rates. Across all sites, more than 3,300 person trips were observed, providing a suitable basis from which an assessment of travel characteristics could be made.

Following the conduct of the multi-modal person trip counts, a vehicle peak hour for each site was determined based on the largest number of vehicle trips that were observed on a weekday (either a Tuesday or Thursday) and Saturday. The number of peak hour vehicle trips was then divided by the number of dwellings to derive a vehicle trip generation rate for each site for an average weekday and Saturday. A comparison of vehicle trip generation rates between case and control

Table 8.3 Total person trips (all transport modes) observed at case and control sites

Site ID	Total dwellings	Total number of person trips observed				Total person trips observed/ dwelling
		Tuesday (7 am–9 am)	Thursday (7 am–9 am)	Saturday (10 am–1 pm)	Total trips observed	
Case-control 1 (CC1)						
Case site	242	232	229	465	926	3.8
Control site	54	52	54	43	149	2.8
Case-control 2 (CC2)						
Case site	282	225	231	457	913	3.2
Control site	156	170	158	280	608	3.9
Case-control 3 (CC3)						
Case site	124	111	122	199	432	3.5
Control site	45	39	30	36	105	2.3
Case-control 4 (CC4)						
Case site	24	31	28	41	100	4.2
Control site	34	31	24	69	124	3.6
Sub-total						
Case sites	672	599	610	1,162	2,371	3.5
Control sites	289	292	266	428	986	3.4
Grand total	961	891	876	1,590	3,357	3.5

sites was then undertaken. In addition, supplementary comparisons were also made to published vehicle trip generation rates. These included:

- Institute of Transportation Engineers (ITE) 2008 rates for high-rise apartments based on data from the United States (Institute of Transportation Engineers 2008)
- Roads and Traffic Authority (RTA) 2002 rate for high density residential flats based on data from the Australian state of New South Wales (Roads and Traffic Authority 2002)
- Roads and Maritime Services (RMS) 2013 rates for high density residential flats based on data from Sydney, Australia (Roads & Maritime Services 2013).

Transport mode shares were also calculated for each site for an average weekday from 7 am–9 am (based on the data collected on the Tuesday and Thursday) and for a Saturday from 10 am–1 pm. The difference in the average transport mode shares between the case and control sites was then calculated. Statistical tests were also conducted to determine whether the differences in mode shares were statistically significant.

In addition, the average car driver mode share observed at each site was compared to regional travel survey data for the relevant local government area of each site (for trips to/from home during the same time periods), as a supplementary point of comparison. The regional travel survey data was based on the 2009–10 Victorian Integrated Survey of Travel and Activity (VISTA), a household travel survey conducted in the state of Victoria (Department of Transport, Planning and Local Infrastructure 2010). While a comparison could have also been undertaken against journey to work trips from the 2011 Census (Australian Bureau of Statistics 2011a), this was not considered to be appropriate given that non-work trips were also observed at the case and control sites during the times at which the multi-modal person trip counts were conducted.

8.3.2.2 Car and Bicycle Parking Utilisation Surveys

A count of the total demand and supply of car and bicycle parking facilities on a Tuesday and Thursday (at 7 am) was conducted at each site, corresponding to the same weekdays in which the multi-modal person trip counts were undertaken. Formal approval to conduct the car and bicycle parking utilisation surveys was sought in a letter to the property manager of each site. The letter was prepared on Monash University letterhead and was signed by the researcher's supervisor. A copy of the letter (with identifying features removed) is provided in Appendix D. Following receipt of the letter, each property manager gave approval to conduct the surveys on the condition that the surveyors would be accompanied by a building manager or caretaker located at the site during working hours.

While it would have been preferable to conduct the parking surveys earlier in the day to correspond with peak times (e.g. before 5 am, due to the residential nature of the sites), this was not possible due to the time at which the building managers and caretakers started their working day. However, as data was collected at a given case site at the same time as its corresponding control site, this was not considered to cause any major comparability issues. It is also noted that the surveys did not include a count of bicycles stored within individual apartments or a count of residents' cars parked outside of the building (i.e. on-street). The surveys therefore only considered on-site car and bicycle parking facilities.

As with the multi-modal person trip counts, the 16 civil engineering undergraduate students supported the resourcing requirement for the car and bicycle parking utilisation surveys as part of their tailored assignment for one of their transport engineering units.

Following the conduct of the car and bicycle parking utilisation surveys, the rate of car and bicycle parking demand at each site was determined. This was equivalent to the largest number of observed cars and bicycles parked on-site (on a Tuesday and Thursday), divided by the total number of dwellings. The number of observed cars and bicycles parked was also divided by the total supply of on-site car and bicycle spaces to determine the level of utilisation of parking facilities at each site. The differences in parking demand and utilisation rates between the case and control sites were then calculated. Statistical tests were conducted to understand the level of statistical significance in the differences that were observed.

8.3.2.3 Resident Travel Survey

A self-completion questionnaire was also developed and administered anonymously to residents living at the case and control sites during May 2014. A total of 19 questions were included that covered trip frequency by mode and purpose, car and bicycle ownership, attitudes and preferences towards different transport modes, and demographics. No mention of the travel plan was made to avoid any potential bias associated with those who might be more likely to respond to the survey because they changed their travel behaviour in accordance with the intentions of the travel plan. However, a question asked respondents about their awareness and use of travel initiatives that may be available at their building, some of which formed part of the travel plan. A copy of the questionnaire is provided in Appendix G.

The survey was hosted online, with a link (URL) developed for each site along with a Quick Response (QR) code for residents wishing to complete the survey using a smart phone. Hard copies of the questionnaire were made available to residents upon request. A prize draw for completing the survey was offered in the form of a $250 retail voucher. This was chosen over the use of multiple vouchers totalling the same amount (e.g. 10 × $25) given that individuals tend to place a greater focus on the value of a prize rather than the probability of winning (Kahneham and Tversky 1979). While the survey was anonymous, respondents had to provide some basic contact details for follow-up purposes (first name only, plus an email address or phone number) if they wished to be eligible for the prize draw.

Formal approval to conduct the travel survey was sought from the property manager of each site as part of the same letter that requested access to conduct the car and bicycle parking utilisation surveys (a copy of the letter is provided in Appendix D). Approval was conditioned on delivering the travel survey material during business hours only due to security access arrangements at most sites. In addition, survey material could only be delivered to each mail box as personal contact (e.g. door knocking) was not permitted by the property managers. The survey material comprised a pre-notification letter, then a survey postcard (containing the URL and QR code) delivered 3 days later, followed by a reminder postcard delivered 1 week after the survey postcard. All survey material was professionally designed and included the University branding. Copies of the pre-notification letter and survey postcard are provided in Appendix E and F respectively. The reminder postcard was the same as the initial survey postcard with the exception of a sticker that was placed on the front side to remind residents of the closure date for the prize draw. Pre-notification letters and survey postcards were placed within individually addressed envelopes to increase the chance they would be opened. Reminder postcards were not placed in envelopes or delivered to mail boxes requesting 'no junk mail' (which comprised only 3.5 % of total mail boxes). Building managers at two sites assisted in further promoting the survey through posters, electronic displays, and an email to residents.

Table 8.4 Resident travel survey responses

Characteristic	Case sites	Control sites	All sites
Total dwellings	672	289	961
Total occupied dwellings[a]	564	237	801
Total complete survey responses			
via hard copy	0	0	0
via URL	60	46	106
via QR code	5	5	10
Total	65	51	116
Survey responses via QR code (%)	7.7	9.8	8.6
Survey response rate (%)	11.5	21.5	14.5

[a]Based on estimates provided by property/building managers at each site

Table 8.4 details the number of survey responses received. A total of 116 complete responses were received from 801 occupied dwellings, providing a response rate of 14.5 %. This is consistent with response rates of 11–20 % achieved in similar studies (Lee et al. 2014). It is also worth comparing this to an 8 % response rate achieved by the Victorian Government in a pilot survey of residents living in high-rise apartment buildings in Melbourne in 2009, where the secure nature of buildings precluded personal contact with respondents (Roddis 2014, personal communication). Accessing secure apartment buildings has been reported as a challenge even when conducting the Australian Census (Australian Bureau of Statistics 2011b) despite resident participation being compulsory under the Census and Statistics Act 1905 (Australian Government 2006).

Table 8.4 shows that a minority (around 9 %) of survey respondents used the QR code while most (91 %) manually typed the URL into a web browser. No hard copies of the questionnaire were requested by residents. Respondents that used the QR code completed the survey in less time (average of 6.8 min) compared to those who manually typed in the URL (average of 9.0 min). This difference in survey completion time was statistically significant ($p = 0.03$) and provides support towards considering the use of QR codes in future travel surveys. While not shown in Table 8.4, sites where building managers helped to promote the survey achieved a higher response rate (16.6 %) compared to those without building management support (12.4 %). This difference in response rates was statistically significant ($p = 0.05$). Most respondents (93 %) provided their contact details to be eligible for the prize draw, thereby indicating the potential importance of this incentive in attracting survey responses.

To assess broad representativeness of the survey sample, Table 8.5 provides a comparison of the sample to the relevant local area populations from the 2011 Census. The local area populations are limited to residents living in apartment buildings of four or more storeys to provide comparability with the survey sample. The average household size and proportion of renting households in the sample was generally representative of the local area population. While greater variation is

Table 8.5 Comparison of resident survey sample and local area population[a] characteristics

Characteristic	South Yarra		South Melbourne		Brunswick	
	Survey sample[b]	Local area population	Survey sample[c]	Local area population	Survey sample[d]	Local area population
Total households	36	3,147	47	1,801	33	913
Average household size	1.59	1.73	1.96	1.84	1.88	1.83
Renting households (%)	86	71	65	65	76	69
Car ownership						
0 car households (%)	50	32	17	30	35	32
1 car households (%)	39	51	59	54	50	51
2 + car households (%)	11	17	24	16	15	17
Average cars/household	0.61	0.85	1.13	0.88	0.82	0.87
Gender						
Male (%)	56	48	46	49	29	53
Female (%)	44	52	54	51	71	47
Age						
20–39 years (%)	89	53	67	55	91	69
40–59 years (%)	11	20	13	20	6	13
60 years or more (%)	0	19	13	15	3	10
Income						
Less than $300 per week (%)	3	21	16	28	0	23
$300–$599 per week (%)	17	16	2	15	13	17
$600–$999 per week (%)	17	19	18	14	22	20
$1,000–$1,499 per week (%)	31	19	31	19	28	23
$1,500 or more per week (%)	33	25	33	24	38	18

[a]*Source* Australian Bureau of Statistics (2011a); local area population figures are limited to apartment buildings of four or more storeys to provide comparability with the survey sample
[b]Corresponds to case-control 1
[c]Corresponds to case-control 2
[d]Corresponds to case-control 3 and 4

observed for other characteristics, this is not of concern given the target population was highly selective by only considering residents living at particular apartment buildings.

Following the conduct of the resident travel survey, the data was analysed using descriptive statistics to understand the level of awareness and use of travel plan measures among respondents living at the case sites. Descriptive statistics were also used to understand reported differences in travel behaviour between respondents living at the case sites and control sites. Statistical tests were then conducted to understand the level of statistical significance in the differences reported by respondents.

Propensity Score Matching (PSM) was then used to assess the extent of self-selection among residents living at the case sites. The general process that was adopted is illustrated in Fig. 8.3.

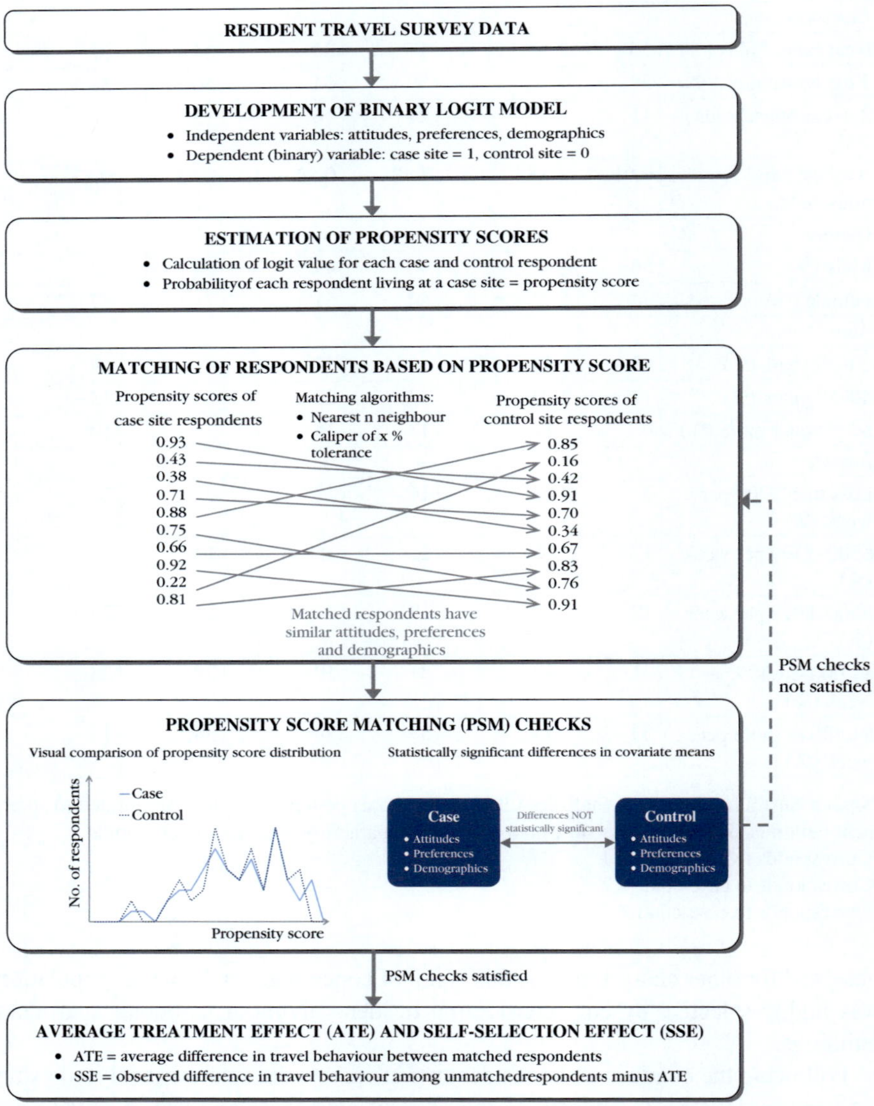

Fig. 8.3 Propensity score matching (PSM) process

The PSM technique involved firstly developing a binary logit model to predict the probability of a resident living at a case site based on a set of attitudinal, preference and demographic variables. These variables were chosen in light of their recognition as key sources of self-selection and their use as controls in self-selection studies (Cao et al. 2010; Mokhtarian and Cao 2008; Naess 2009).

To predict the probability of a resident living at a case site, a logit value was first required for each respondent. This was calculated as the sum of each binary logit model coefficient multiplied by the value of the respective independent variable for each respondent, as follows:

$$logit_i = \beta_0 + \beta_1 x_{i1} + \beta_2 x_{i2} + \cdots + \beta_k x_{ik} \tag{8.1}$$

The odds of a respondent living at a case site could then be calculated using the logit value:

$$odds_i = e^{logit_i} \tag{8.2}$$

The probability of a respondent living at a case site, denoted as PS_i, also known as the propensity score which ranges from 0 to 1, could then be calculated based on the odds:

$$PS_i = \frac{odds_i}{1 + odds_i} \tag{8.3}$$

Each case site respondent was then matched with a control site respondent based on having a similar propensity score. Matched respondents therefore had similar attitudes, preferences and demographics, and therefore a similar propensity to live at a case site. In accordance with PSM guidance, matching was based on the log odds of the propensity score so that the density of scores would be well spread and a consistent bandwidth could be used (Heinrich et al. 2010).

The matching process was carried out using Stata 13, a statistical software package. Previous studies have opted for a relatively high level of tolerance in the matching process by specifying a maximum difference in the propensity score, referred to as the 'caliper', of 0.01 (Cao et al. 2010; Lee et al. 2014). Therefore, a caliper of 0.01 was initially adopted, although a caliper of 0.02 was also tested. However, both of these methods resulted in a significant number of unmatched respondents being discarded from the process given the relatively small sample available. Matching was therefore undertaken at a lower tolerance level using the 'nearest five neighbours' algorithm, in accordance with guidance on PSM (Heinrich et al. 2010). This involves matching case site respondents with the weighted average of the nearest five control site respondents. The benefit of using this method in the case of small samples is generally considered to outweigh the impact of any sampling error that is introduced (Heinrich et al. 2010). Matching was also undertaken using the nearest three neighbours and the nearest four neighbours. However, the nearest five neighbours was found to be most appropriate with the sample size that was available.

To ensure the matching process was adequate, a number of checks were undertaken in line with PSM guidance (Caliendo and Kopeinig 2008; Heinrich et al. 2010).

Firstly, visual comparisons of the propensity score distribution between case and control site respondents were undertaken to ensure there was sufficient overlap between the distributions. Secondly, differences in covariate (independent variable) means between case and control site respondents were tested for statistical significance to ensure they were adequately balanced between the two groups.

The average difference in travel behaviour between matched respondents was then calculated to represent the Average Treatment Effect (ATE), or in other words, the effect of the travel plan after self-selection is taken into account. The Self-Selection Effect (SSE) was then calculated as the average of the Observed Difference (OD) in travel behaviour among *unmatched* respondents, minus the Average Treatment Effect (ATE).

8.4 Results

This section presents the results associated with the data collection and analysis undertaken for each site. This includes an assessment of:

- Vehicle trip generation rates (based on the multi-modal person trip count data)
- Transport mode shares (based on the multi-modal person trip count data)
- Car and bicycle parking utilisation (based on the parking utilisation survey data)
- Awareness and use of travel plan measures (based on the resident travel survey data)
- Residential self-selection (based on the resident travel survey data).

8.4.1 Vehicle Trip Generation Rates

The vehicle trip generation rate observed at each case and control site is shown in Fig. 8.4 (average weekday AM peak hour) and Fig. 8.5 (Saturday peak hour). For comparative purposes, published vehicle trip generation rates available from the Institute of Transportation Engineers (ITE), Roads and Traffic Authority (RTA) and Roads and Maritime Services (RMS) are also shown.

With the exception of the Saturday peak hour vehicle trip generation rate for case site 1, all case sites had a lower vehicle trip generation rate both on weekdays and Saturdays when compared to their corresponding control sites. In addition, the published vehicle trip generation rates are not only higher than each case site, but also higher than almost all control sites. Therefore, the difference in vehicle trip generation rates would have been grossly overestimated if published rates had been assumed to apply in the absence of any control sites.

There are a number of reasons why the published vehicle trip generation rates are mostly higher than those at the case and control sites. Firstly, while the type of dwelling (in this case, high density apartments) is accounted for, the published

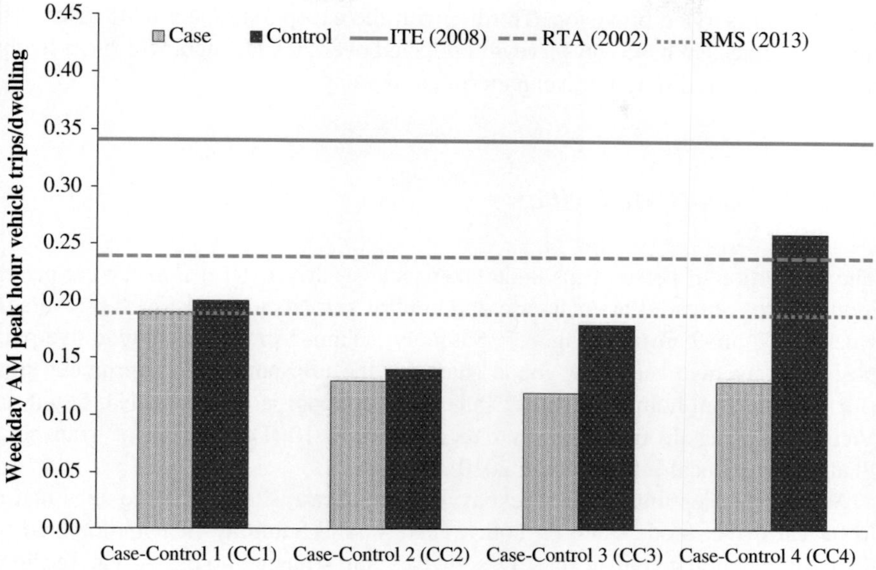

Fig. 8.4 Comparison of vehicle trip generation rates—average weekday AM peak hour

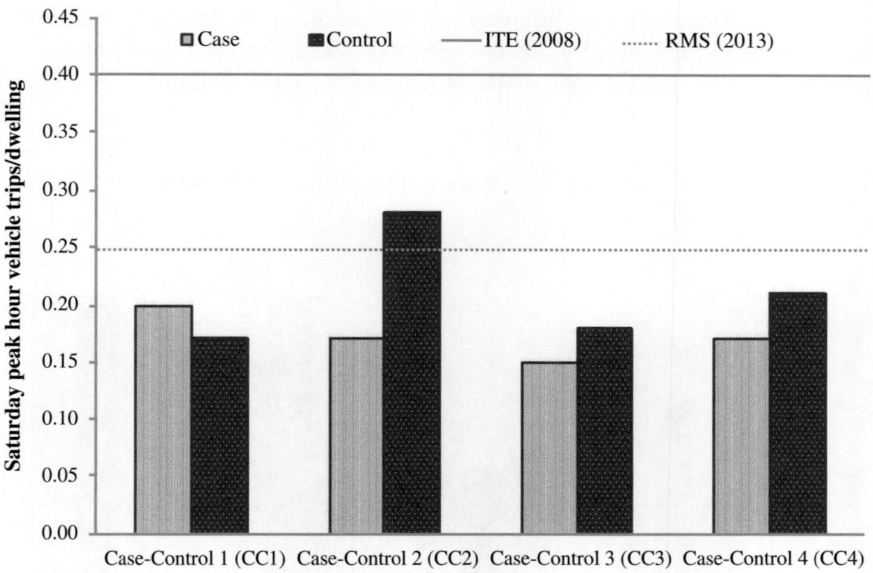

Fig. 8.5 Comparison of vehicle trip generation rates—Saturday peak hour

rates are based on survey data from a very limited sample of sites (as little as five sites for the ITE Saturday peak hour rate) and therefore may not be representative. Secondly, the published rates are based on sites from different locations than the case and control sites and are therefore unlikely to have a similar level of transport network and service provision. Thirdly, with the exception of the RMS rates, the published rates are based on surveys conducted over 20 years ago, which are likely to have observed different travel patterns from today.

8.4.2 Transport Mode Shares

The proportion of person trips undertaken as a car driver (termed as the car driver mode share) observed at each case and control site is shown in Fig. 8.6 (average weekday, 7 am–9 am) and Fig. 8.7 (Saturday, 10 am–1 pm). For comparative purposes, the average car driver mode share for the relevant local government area (for trips to/from home during the same time periods) is also provided, based on Victorian household travel survey data from 2009–10 (Department of Transport, Planning and Local Infrastructure 2010).

With the exception of the weekday pattern at case site 2, all case sites had a lower car driver mode share on both weekdays and Saturdays when compared to their corresponding control sites. Despite case site 4 having no on-site car parking, car driver trips were still made to and from this site through the use of on-street car parking. Also of note is that the average car driver mode shares across the local government areas are not only higher than each case site, but also higher than each control site. Therefore, the difference in the car driver mode share would have been grossly overestimated if the average mode shares for each local government area had been assumed to apply in the absence of any control sites.

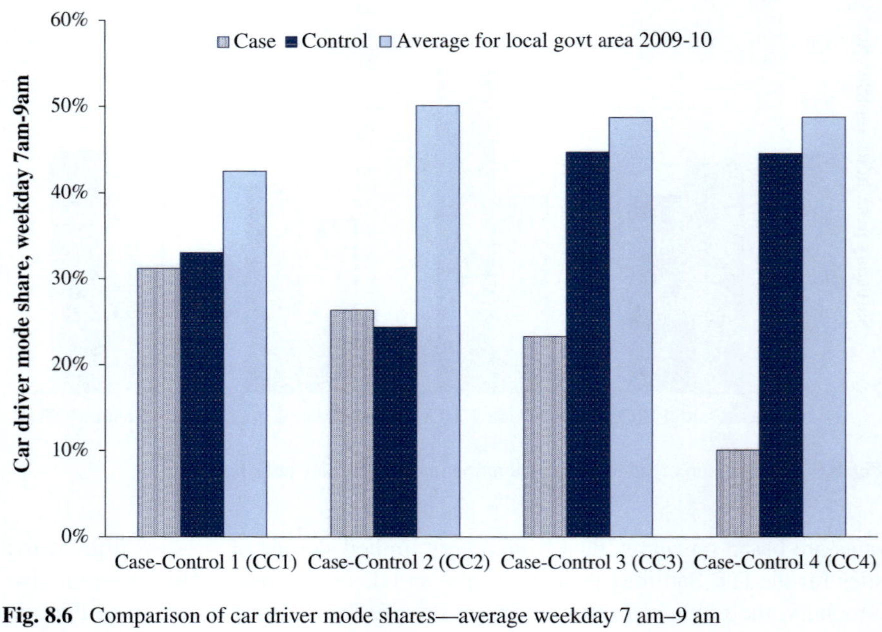

Fig. 8.6 Comparison of car driver mode shares—average weekday 7 am–9 am

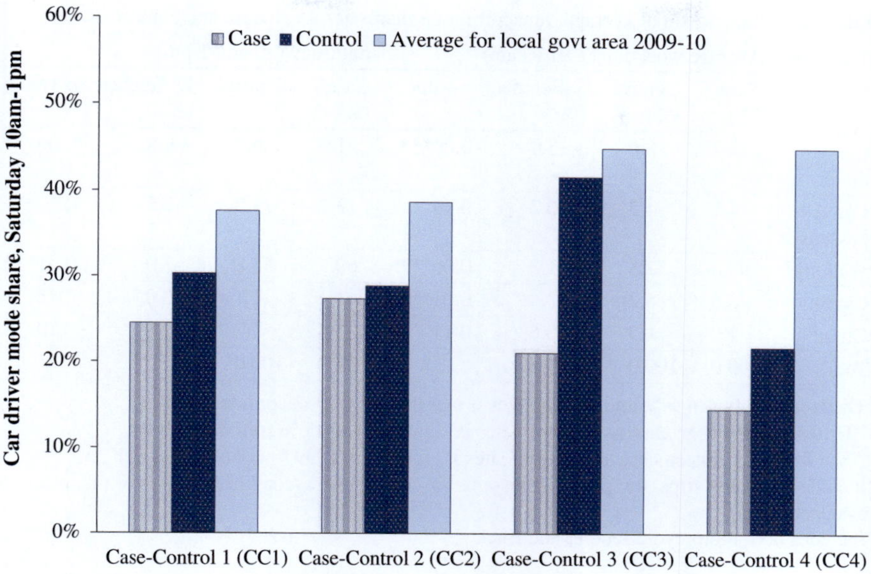

Fig. 8.7 Comparison of car driver mode shares—Saturday 10 am–1 pm

There are a number of reasons for the higher car driver mode share average for each local government area. Firstly, the local government area averages are based on household travel survey data which accounts for all types of housing (not just apartments) and is therefore likely to encompass different rates of car parking, as well as variation in terms of the socio-demographic characteristics of residents, when compared to the case and control sites. Secondly, the specific location of the case and control sites may not be representative of average transport network and service provision for the entire local government area. Thirdly, the household travel survey data was collected in 2009–10 while the data for the case and control sites was collected in 2014. Since 2009–10, general improvements to public transport, walking and cycling infrastructure may have influenced car use.

A comparison of average transport mode shares across all case and control sites is provided in Table 8.6. On an average weekday (7 am–9 am), the average car driver mode share was 14 % points lower at the case sites than the control sites. Furthermore, the average mode share for walking was 11 % points higher at the case sites than the control sites, while the average mode share for cycling was 3 % points higher. All of these differences were statistically significant (p-values ranging from 0.00 to 0.01). On a Saturday (10 am–1 pm), the average mode share for car driver trips was 9 % points lower at the case sites, while the average mode share for walking was 7 % points higher. Both of these differences were also statistically significant (p-values ranging from 0.00 to 0.01). No difference in the average mode share for cycling on a Saturday was observed between the case and control sites. As the counts were only conducted at the place of residence, public

Table 8.6 Comparison of average transport mode shares across all case and control sites

Transport mode	Average weekday (7 am–9 am)				Saturday (10 am–1 pm)			
	Case (%)	Control (%)	Difference (%)	p-value	Case (%)	Control (%)	Difference (%)	p-value
Car as driver	22.7	36.6	–13.9	0.00***	21.9	30.7	–8.8	0.00***
Car as passenger	4.1	4.2	–0.2	0.46	13.7	11.2	+2.5	0.10*
Walking[a]	63.6	52.3	+11.3	0.00***	60.9	53.8	+7.0	0.01***
Cycling	8.6	5.2	+3.4	0.01***	1.1	1.1	0.0	0.48
Other[b]	1.0	1.7	–0.6	0.11	2.4	3.2	–0.8	0.19
Total	100.0	100.0			100.0	100.0		

*Difference between case and control sites is significant at 90 % confidence level
**Difference between case and control sites is significant at 95 % confidence level
***Difference between case and control sites is significant at 99 % confidence level
[a]Includes onward trips by public transport as counts were conducted only at the place of residence
[b]Includes taxi, motorcycle/scooter and truck

transport trips could not be observed. However, given the proximity of the sites to public transport, it is likely that the mode share for walking would have incorporated most onward trips made by public transport.

8.4.3 Car and Bicycle Parking Utilisation

Table 8.7 provides a comparison of on-site car and bicycle parking demand at the case and control sites. On average, the control sites exhibited a higher level of car parking demand and a lower level of bicycle parking demand than the case sites (statistically significant, p-value = 0.00). At an individual paired case-control level, while bicycle parking demand was consistently lower across all control sites, car parking demand was also lower at two of the control sites (1 and 2).

Table 8.8 presents a comparison of on-site car and bicycle parking utilisation at the case and control sites. On average, the case sites had a higher rate of both car and bicycle parking utilisation, reflecting a greater level of space efficiency (statistically significant, p-values = 0.00 and 0.01). Table 8.8 also shows (for both case and control sites) that bicycle parking is in most cases close to or over capacity while car parking is mostly under capacity. This may lead to some residents having to store their bicycle/s within their apartment and not in the common bicycle storage area, meaning that the actual number of bicycles per dwelling could be higher than that shown in Table 8.7.

Table 8.7 Comparison of on-site car and bicycle parking demand at case and control sites

Site ID	Cars parked per dwelling				Bicycles parked per dwelling			
	Case	Control	Difference	p-value	Case	Control	Difference	p-value
Case-Control 1 (CC1)	0.40	0.37	0.03	0.34	0.33	0.30	0.03	0.33
Case-Control 2 (CC2)	0.75	0.62	0.13	0.00***	0.44	0.39	0.05	0.16
Case-Control 3 (CC3)	0.51	0.62	−0.11	0.10	0.69	0.49	0.20	0.01**
Case-Control 4 (CC4)	0.00[a]	0.59	−0.59	0.00***	1.46	0.56	0.90	0.00***
Average	0.42	0.55	−0.13	0.00**	0.73	0.43	0.30	0.00***

*Difference between case and control site is significant at 90 % confidence level
**Difference between case and control site is significant at 95 % confidence level
***Difference between case and control site is significant at 99 % confidence level
[a]No car parking is provided at case site 4

Table 8.8 Comparison of on-site car and bicycle parking utilisation at case and control sites

Site ID	Car parking utilisation				Bicycle parking utilisation			
	Case (%)	Control (%)	Difference (%)	p-value	Case (%)	Control (%)	Difference (%)	p-value
Case-Control 1 (CC1)	59	47	12	0.08*	120	94	26	0.00***
Case-Control 2 (CC2)	79	61	18	0.00***	155	47	108	0.00***
Case-Control 3 (CC3)	65	49	16	0.03**	77	122	−45	0.00***
Case-Control 4 (CC4)	n.a.[a]	50	n.a.[a]	n.a.[a]	56	127	−71	0.00***
Average	68	52	16	0.00***	102	98	4	0.01**

*Difference between case and control site is significant at 90 % confidence level
**Difference between case and control site is significant at 95 % confidence level
***Difference between case and control site is significant at 99 % confidence level
[a]No car parking is provided at case site 4

8.4.4 Awareness and Use of Travel Plan Measures

Residents were asked about their level of awareness and use of various travel initiatives that may be available at their apartment building. Table 8.9 provides a summary of the results. With the exception of transport information in the building users guide, the common bicycle storage area, and car sharing vehicles on the street, use of existing travel initiatives was relatively low across all sites (around

Table 8.9 Awareness and use of travel initiatives among survey respondents

Travel initiative	Aware, used (%)	Aware, not used (%)	Not aware (%)
Case sites			
Transport information in building users guide[a]	60	40	0
Common bicycle storage area	51	43	6
Car sharing vehicle on street[a]	40	60	0
Free membership to car share	11	45	45
Car sharing vehicle/s in building car park	9	53	38
Transport information in new residents kit	9	14	77
Transport information on building's website	7	13	80
Display in lobby containing transport information[a]	7	7	87
Free public transport tickets for new residents	7	0	93
Bicycle fleet in building car park[a]	4	32	64
Online resident forum for organising carpooling[a]	4	21	75
Common shower facilities[a]	0	100	0
Free membership to Melbourne Bike Share[a]	0	53	47
Umbrellas at reception/lobby area	0	7	93
Control sites			
Common bicycle storage area	37	43	20
Transport information on building's website	3	0	97

Note percentages are based only on sites where respective initiatives are in place
[a]Initiative offered only at one site

10 % or less). Awareness of existing travel initiatives was also low in some cases, particularly initiatives relating to transport information on the building's website, free public transport tickets for new residents, and umbrellas at the reception/ lobby area. However, it is noted that use of the common bicycle storage area is greater at the case sites (51 %) than the control sites (37 %). This difference was found to be statistically significant at the 90 % confidence level ($p = 0.07$).

8.4.5 Residential Self-selection

Using the results from the resident travel survey, Table 8.10 provides a summary of travel behaviour characteristics, in addition to travel related attitudes and preferences, among respondents residing at the case and control sites.

Table 8.10 Characteristics of survey respondents residing at the case and control sites

Characteristic	Case sites (n = 65)	Control sites (n = 51)	Difference	p-value	Statistical power[a] (%)
Frequency of using transport mode by trip purpose[b]					
Work—public transport	3.16	2.98	0.18	0.33	19.7
Work—walk	1.00	1.19	−0.19	0.29	22.9
Work—car driver	1.06	1.52	−0.46	0.12	45.6
Work—car passenger	0.50	0.36	0.14	0.21	32.6
Work—bicycle	0.88	0.38	0.50	0.04**	70.3
Shopping—public transport	1.65	1.71	−0.06	0.42	13.9
Shopping—walk	2.98	2.41	0.57	0.04**	67.8
Shopping—car driver	1.40	2.00	−0.60	0.04**	68.9
Shopping—car passenger	1.03	1.14	−0.11	0.34	19.0
Shopping—bicycle	0.74	0.31	0.42	0.02**	78.1
Car ownership					
0 car households (%)	35.4	29.4	6.0	0.25	27.3
1 car households (%)	50.8	49.1	1.7	0.43	13.6
2 + car households (%)	13.8	21.5	−7.7	0.14	42.6
Average cars/household	0.78	1.00	−0.22	0.06*	58.3
Bicycle ownership					
0 bicycle households (%)	41.6	43.2	−1.6	0.43	13.4
1 bicycle households (%)	24.6	29.4	−4.8	0.28	24.3
2 + bicycle households (%)	33.8	27.4	6.4	0.23	29.2
Average bicycles/household	1.02	0.88	0.14	0.22	30.8
Attitudes and preferences[c]					
I prefer to travel by car—whenever possible	2.80	2.82	−0.02	0.92	11.6
It is important that I have my own allocated car parking space at home	3.51	4.08	−0.57	0.03**	83.3
Others may think I had a financial difficulty if I did not have a car	1.66	1.78	−0.12	0.49	27.5
I prefer to take public transport than travel by car—whenever possible	3.51	3.49	0.02	0.94	11.7

(continued)

Table 8.10 (continued)

Characteristic	Case sites (n = 65)	Control sites (n = 51)	Difference	p-value	Statistical power[a] (%)
I prefer to use a bicycle than travel by car—whenever possible	2.74	2.59	0.15	0.53	25.8
I like to live where I have shops within walking distance of my home	4.49	4.47	0.02	0.87	13.0

[a]Acceptable level of statistical power is considered to be 80 % (Cohen 1988)
[b]0 = never (0 % of the time) to 5 = almost always (80–100 % of the time)
[c]1 = strongly disagree to 5 = strongly agree
*Significant at 90 % confidence; **Significant at 95 % confidence

Consistent with the multi-modal person trip count data, case site respondents reported a lower frequency of car driver trips than control site respondents. Case site respondents also reported a mostly higher frequency of trips by non-car driver modes (public transport, walk, car passenger, and bicycle) than control site respondents. While differences in travel frequencies by mode between case and control site respondents were observed, these were only statistically significant in some cases (bicycle trips for work; walk, car driver and bicycle trips for shopping). This is likely to be due to the relatively small sample size given the statistically significant differences that were observed in transport mode shares through the multi-modal person trip counts.

Consistent with the car parking utilisation surveys, case site respondents reported lower car ownership (0.78 cars per household) compared to control site respondents (1.00 cars per household). This difference was statistically significant ($p = 0.06$). A number of comments were made by case site respondents that supported the concept of lower car ownership:

> Moving here has been life changing. I sold my car with glee - keen for the savings and the lifestyle change…Even for my son - walking or riding to school with him has changed the way we interact - significantly improving the quality of our time together - with a lot less rushing - which is good for everyone I believe [Case site resident, female 30 to 39 years old].

> I am more than happy to go without a vehicle in exchange for fairly priced public transport options. I utilise Go Get [car sharing] extensively and have no desire to ever own a vehicle again [Case site resident, male 30 to 39 years old].

On average, case site respondents reported higher bicycle ownership (1.02 bicycles per household) compared to control site respondents (0.88 bicycles per household), however this difference was not statistically significant ($p = 0.22$). The lack of statistical significance is likely to be due to the limited sample size of the resident travel survey given that the bicycle parking utilisation surveys showed a statistically significant difference in average bicycle parking demand between the case and control sites.

Very little difference in travel related attitudes and preferences were reported by case and control site respondents. This may suggest that a limited amount of self-selection among case site residents has occurred given some statistically significant differences in travel characteristics between case and control site respondents. However, control site respondents were more likely to agree that having an allocated car parking space is important (agreement rating of 4.08 vs. 3.51 out of 5), with this difference being statistically significant ($p = 0.03$).

Despite the presence of some statistically significant differences in reported travel characteristics among case and control site respondents, the level of 'statistical power' based on the t-tests undertaken was insufficient to detect a difference due to the limited sample size available. Cohen (1988) suggests that an acceptable level of statistical power is 80 %. However, as shown in Table 8.10, the level of statistical power achieved is lower than 80 % for all variables except one (attitudes relating to car parking). This calls for caution when interpreting the results.

Table 8.11 presents the binary logit model which was used to estimate the probability of a resident living at a case site given a set of attitudinal, preference and demographic variables. As the logit model is a prediction model used only to extract the propensity score, the statistical significance of independent variables and any multicollinearity is not a concern (Cao et al. 2010).

Table 8.11 Binary logit model for the choice of residing at a case site (vs. a control site)

Independent variable (covariate)	Coefficient	Standard error	p-value
Constant	3.22	3.26	0.32
Attitudes and preferences			
Prefer to travel by car	0.00	0.27	0.99
Important to have car parking space	−0.21	0.19	0.28
Financial difficulty without car	−0.14	0.23	0.54
Prefer to take public transport	−0.08	0.27	0.78
Prefer to use bicycle	0.13	0.19	0.49
Like shops within walking distance	−0.22	0.34	0.52
Demographics			
Household size	−1.11	0.40	0.01***
Years living at current residence	1.22	0.46	0.01***
Housing tenure	−0.39	0.54	0.47
Employment status	−0.50	0.48	0.30
Education status	0.45	0.43	0.29
Income	0.52	0.37	0.15
Age	−0.10	0.24	0.69
Gender	0.27	0.46	0.57

$N = 116$
Log-likelihood at zero = −79.56
Log-likelihood at convergence = −65.82
Pseudo R-square = 0.17
***Significant at 99 % confidence

The propensity score matching process was carried out using Stata 13. To ensure the matching process was adequate, a comparison of the propensity score distribution for case and control site respondents is shown in Fig. 8.8. As can be seen, the distributions are relatively similar after matching with the ‘nearest five neighbours’ algorithm. While matching at a higher level of tolerance, such as a 0.01 or 0.02 caliper, can further improve the level of agreement between the distributions, as illustrated by Fig. 8.8c, d, a significant and unacceptable loss in the sample occurs as many respondents cannot be matched.

A comparison of the differences in covariate (independent variable) means between case and control site respondents before and after matching was also undertaken, as shown in Table 8.12. Prior to matching, the difference in means is significant for three covariates (important to have car parking space, household size, years living at current residence). However, no significant differences exist after matching. This provides evidence that respondents have been matched appropriately based on a sufficiently similar set of attitudes, preferences and demographics.

Following the matching of case and control respondents, the Average Treatment Effect (ATE) and Self-Selection Effect (SSE) could be calculated. The relative contributions of the travel plan and self-selection was estimated as proportions of the Observed Difference (OD) in travel behaviour, as shown in Table 8.13. Here, transport modes are only included where significant differences in their trip frequencies between case and control respondents were reported, despite the lack

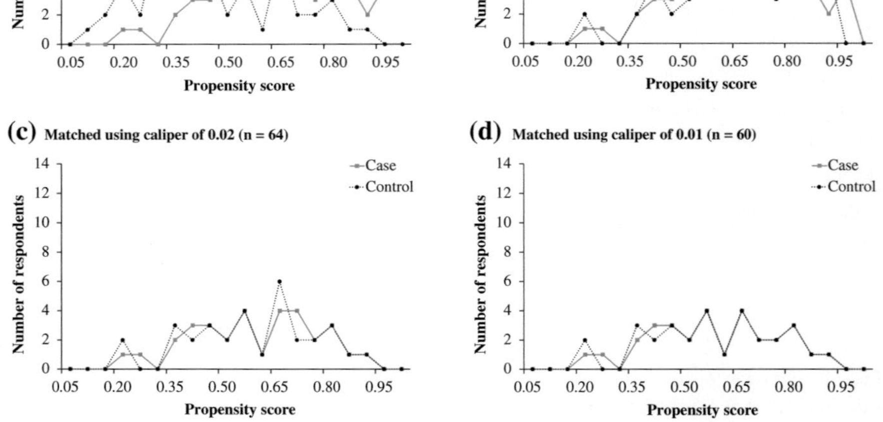

Fig. 8.8 Distribution of propensity scores for the case and control sites under different matching algorithms

Table 8.12 Comparison of covariate means between case and control site respondents before and after matching

Independent variable (covariate)	Unmatched means				Matched means			
	Case sites	Control sites	Difference	p-value	Case sites	Control sites	Difference	p-value
Attitudes and preferences[a]								
Prefer to travel by car	2.80	2.82	−0.02	0.92	2.84	2.85	−0.01	0.94
Important to have car parking space	3.51	4.08	−0.57	0.03**	3.71	3.79	−0.08	0.76
Financial difficulty without car	1.66	1.78	−0.12	0.49	1.73	1.85	−0.12	0.52
Prefer to take public transport	3.51	3.49	0.02	0.94	3.58	3.68	−0.10	0.62
Prefer to use bicycle	2.74	2.59	0.15	0.53	2.73	2.59	0.14	0.56
Like shops within walking distance	4.49	4.47	0.02	0.87	4.42	4.35	0.07	0.65
Demographics								
Household size (no. people)	1.66	2.00	−0.34	0.01***	1.71	1.65	0.06	0.57
Years living at current residence (no. years)	0.96	0.62	0.34	0.01**	0.75	0.68	0.07	0.59
Housing tenure[b]	0.72	0.76	−0.04	0.62	0.71	0.71	0.00	1.00
Employment status[c]	1.43	1.49	−0.06	0.70	1.51	1.43	0.08	0.62
Education status[d]	0.51	0.31	0.19	0.18	0.36	0.51	−0.15	0.34
Income[e]	2.08	1.94	0.14	0.42	2.13	2.05	0.08	0.64
Age[f]	2.82	2.84	−0.03	0.90	2.85	2.89	−0.04	0.87
Gender[g]	0.57	0.55	0.02	0.83	0.56	0.55	0.01	0.91

[a] 1 = strongly disagree to 5 = strongly agree
[b] 0 = not renting, 1 = renting
[c] 0 = not employed, 1 = casual/part-time, 2 = full-time
[d] 0 = not attending, 1 = part-time, 2 = full-time
[e] 0 = negative/nil income, 1 = $100–$799 per week, 2 = $800–$1499 per week, 3 = $1500 + per week
[f] 1 = 19 years or less, 2 = 20 to 29 years, 3 = 30 to 39 years, 4 = 40 to 49 years, 5 = 50 to 59 years, 6 = 60 to 69 years, 7 = 70 years or more
[g] 0 = male, 1 = female
**Significant at 95 % confidence; ***Significant at 99 % confidence

Table 8.13 Relative contributions of the travel plan and self-selection on travel behaviour

Frequency of using transport mode by trip purpose[a]	Observed difference (OD)[b]	Average treatment effect (ATE)[c]	Self-selection effect (SSE)[d]	Contribution of travel plan (ATE/OD) (%)	Contribution of self-selection (SSE/OD) (%)
Work—car driver	−0.46	−0.35	−0.11	76	24
Work—bicycle	0.50**	0.45*	0.05	90	10
Shopping—walk	0.57**	0.48	0.09	84	16
Shopping—car driver	−0.60**	−0.35	−0.25	58	42
Shopping—bicycle	0.42**	0.32*	0.10	76	24

[a]0 = never (0 % of the time) to 5 = almost always (80–100 % of the time)
[b]Represents unmatched difference in frequency of using transport mode between case and control respondents
[c]Represents matched difference in frequency of using transport mode between case and control respondents
[d]Equal to Observed Difference (OD) minus Average Treatment Effect (ATE)
*Significant at 90 % confidence; **Significant at 95 % confidence

of statistical power (refer to Table 8.10). While the difference in work trip frequency as a car driver was not strictly significant ($p = 0.12$), it is still included in Table 8.13 given that reducing the use of this transport mode is a key focus of travel plans.

Table 8.13 shows that, depending on the transport mode and trip purpose, self-selection has *potentially* contributed 10–42 % of the observed difference in travel behaviour between the case and control sites. This implies that the travel plans have still made a contribution to the difference in travel behaviour (in the order of 58–90 %) after accounting for self-selection. However, as indicated in Table 8.13, the ATE results are statistically significant for bicycle trips only. Furthermore, the lack of statistical power presented earlier means that it is not possible to be confident about the results in the absence of a larger sample size. However, the findings demonstrate that it is possible to quantify the extent of self-selection associated with travel plans for new residential developments, given a sufficient sample size.

8.5 Discussion

The results have shown that the average weekday (7 am–9 am) mode share for car driver trips was 14 % points lower at residential developments with travel plans (case sites) compared to similar residential developments without travel plans (control sites). This result is consistent with other studies that have assessed the effectiveness of school and workplace travel plans. For example, Cairns et al.

(2004) found an average reduction in car driver trips of at least 14 % points from workplace travel plans in the United Kingdom, while Howlett and Watson (2010) showed reductions of around 10–15 % points from school and workplace travel plans implemented in Victoria, Australia. Through a comprehensive review of travel planning impacts, Enoch (2012a) found that 'car trips can be cut by up to 30 % in exceptional circumstances, but that a reduction of between 5 and 15 % is more usual at the site/organisation level'.

Key to the success of travel plans is the degree to which specific initiatives are tailored to the needs of the site and its users (Cairns et al. 2004). However, due to the relatively small number of residential developments evaluated as part of this research, it is difficult to infer why some case sites experienced lower levels of car use than others, relative to their corresponding control sites. Despite this, one clear reason that did stand out was the level of car parking provision. Case sites 1 and 2 had a very similar level of car parking provision to their corresponding control sites (generally within 10 %), yet overall, car use at these case sites was only marginally lower than their control sites. In contrast, the rate of car parking provision at case site 3 was 40 % less than its control site (as part of its travel plan) and the difference in car use was significant (around 20 % points). Furthermore, no car parking was provided at case site 4 (as part of its travel plan) and the difference in car use was also significant (up to 35 % points on weekdays). This result is consistent with that of Cairns et al. (2004) who found that workplaces who addressed car parking as part of their travel plan achieved a reduction in car trips of more than 24 % compared to only 10 % for those not addressing this aspect.

Results from the car and bicycle parking utilisation surveys showed a lower level of car parking demand and a higher level of bicycle parking demand at the case sites when compared to the control sites. These findings are consistent with the aim of travel plans in supporting more sustainable travel (Enoch 2012a; Rye 2002b). However, bicycle parking across all sites was generally close to or over capacity, while car parking was generally well under capacity. This finding has important implications, not only for government in stipulating requirements for parking provision at residential developments, but also for property developers given the relative costs of providing car parking versus bicycle parking.

Previous studies have reported reductions in car use associated with travel plans for new residential developments, yet these have generally been based on comparisons to secondary data sources (Arlington County Commuter Services 2013; BioRegional 2009; Department for Transport 2005; WSP 2014). The findings reported in this chapter have shown that in almost all cases, the average level of car use based on secondary data sources was not only higher than the case sites but also higher than the control sites. Key reasons for this were due to inconsistencies between the secondary data and the case/control site data, particularly in terms of their geographical locations, housing types and data collection periods. Previous studies that have only made comparisons to secondary data sources may therefore be overestimating the impact of travel plan interventions. Even in the absence of a travel plan, differences in published vehicle trip generation rates of more than 40 % have been observed at transit oriented developments (Arrington and Cervero 2008). The findings therefore clearly illustrate the value of using control sites when evaluating the impacts of travel plans at new residential developments. However, in doing so, the practicality of finding suitable control sites and also resourcing the collection of data at these sites needs to be considered.

The results from the resident travel survey indicated a relatively low level of use, and in some cases, awareness of travel initiatives available at the case and control sites. Consistent with best practice guidance on residential travel planning (Department for Transport 2005), this highlights the need for renewed efforts to be directed towards actively promoting travel plans where they have been introduced, not only upon occupation, but also on an ongoing basis to ensure awareness levels are maintained. This is particularly relevant given the continual turnover of new residents over time. Opportunities for enhancing the implementation of travel plans for new residential developments are discussed further in Chap. 9.

In applying Propensity Score Matching (PSM) to the resident survey data, it was estimated that residential self-selection *potentially* accounts for 10–24 % of the observed difference in travel behaviour between case and control site residents. However, in the case of shopping trips undertaken as a car driver, the contribution of self-selection is potentially much greater at 42 %. Coincidentally, this full range of 10–42 % matches a previous review of 38 studies by Cao et al. (2009) which also found that self-selection contributed 10–42 % of the variation in travel behaviour, albeit in the context of the built environment. Given the potential for residential self-selection at sites with travel plans, future housing and planning policies should give consideration to these preferences where appropriate.

Finally, it is worth noting two key limitations of the findings reported in this chapter. Firstly, only four residential developments with travel plans were evaluated due to resource limitations. Ideally, more sites would help to establish a stronger base evidence base regarding residential travel plan effectiveness, particularly in determining the relative effectiveness of different travel plan measures. Secondly, despite best efforts to attract a sufficient response rate, the sample size available from the resident travel survey limited the ability to make any firm judgements about the extent of residential self-selection due to a lack of statistical power. Despite this, the findings demonstrate the ability to quantify the extent of self-selection associated with travel plans for new residential developments.

8.6 Conclusion

The aim of this chapter is to evaluate the effectiveness of travel plans for new residential developments. In doing so, a range of data collection techniques were undertaken at four case sites (new residential developments with travel plans) and four matching control sites (similar residential developments without travel plans) in Melbourne, Australia. While a lack of statistical power limited the ability to confidently measure the extent of self-selection associated with travel plans for new residential developments, observational counts and surveys revealed a statistically significant reduction in observed car use at the case sites when compared to the control sites.

The findings have highlighted a number of important implications for evaluating the impacts of travel plans for new residential developments. Firstly, the use of suitable control sites can help to overcome comparability issues associated with secondary data sources. Secondly, the potential for residential self-selection demonstrates the importance of controlling for this phenomenon to avoid overstating the impacts of travel plans.

The next chapter takes the findings from this chapter, along with Chaps. 5–7, to identify opportunities to enhance the implementation and subsequent impacts of travel plans for new residential developments. This is achieved through the application and integration of implementation and planning enforcement theories.

References

Arlington County Commuter Services. (2013). *Residential building transportation performance monitoring study*. Virginia, US: Arlington.

Arrington, G., & Cervero, R. (2008). *TCRaP report 128: Effects of TOD on housing, parking and travel.* Washington, D.C., US: Transportation Research Board of the National Academies.

Australian Bureau of Statistics. (2011). *Census of population and housing data*. Retrieved September 25, 2014 from http://www.abs.gov.au/census.

Australian Bureau of Statistics. (2011b). *Reflecting a nation: Stories from the 2011 Census*. Australia: Canberra.

Australian Government. (2006). *Census and statistics act 1905*. Australia: Canberra.

BioRegional. (2009). *BedZED seven years on: The impact of the UK's best known eco-village and its residents*. UK.

Cairns, S., Sloman, L., Newson, C., Anable, J., Kirkbride, A., & Goodwin, P. (2004). *Smarter choices—changing the way we travel*. UK: Department for Transport.

Caliendo, M., & Kopeinig, S. (2008). Some practical guidance for the implementation of propensity score matching. *Journal of Economic Surveys, 22*(1), 31–72.

Cao, X., Mokhtarian, P., & Handy, S. (2009). Examining the impacts of residential self-selection on travel behaviour: A focus on empirical findings. *Transport Reviews, 29*(3), 359–395.

Cao, X., Xu, Z., & Fan, Y. (2010). Exploring the connections among residential location, self-selection, and driving: Propensity score matching with multiple treatments. *Transportation Research Part A, 44*(10), 797–805.

Cohen, J. (1988). *Statistical power analysis for the behavioral sciences* (2nd ed.). Hillsdale, New Jersey, US: Lawrence Erlbaum Associates.

Department for Transport. (2005). *Making residential travel plans work: Guidelines for new development*. London, UK.

Department of Transport, Planning and Local Infrastructure. (2010). *Victorian integrated survey of travel and activity data*. Retrieved September 25, 2014, from http://www.dtpli.vic.gov.au/.

De Gruyter, C., Rose, G., & Currie, G. (In Press). 'Travel Plans for New Residential Developments: Measuring Self-Selection Effects to Better Understand Travel Behaviour Impacts'. *Transportation Research Record: Journal of the Transportation Research Board*, No. 2564.

De Gruyter, C., Rose, G., & Currie, G. (2015). 'Understanding Travel Plan Effectiveness for New Residential Developments'. *Transportation Research Record: Journal of the Transportation Research Board*, No. 2537, 126–136.

Enoch, M. (2012). *Sustainable transport, mobility management and travel plans*. Surrey, England: Ashgate Publishing Limited.

Heinrich, C., Maffioli, A., & Vazquez, G. (2010). *A primer for applying propensity-score matching: Impact-evaluation guidelines*, Technical Notes No. IDB-TN-161, Inter-American Development Bank.

Institute of Transportation Engineers. (2008). *Trip generation* (8th ed.). US: Washington D.C.

Kahneham, D., & Tversky, A. (1979). Prospect theory: An analysis of decision under risk. *Econometrica, 47*(2), 263–291.

Lee, J. S., Zegras, C., Ben-Joseph, E., & Park, S. (2014). Does urban living influence baby boomer' travel behaviour? *Journal of Transport Geography, 35*, 21–29.

Melia, S. (2009). *Potential for carfree development in the UK*. PhD thesis, University of West England.

Melia, S., Barton, H., & Parkhurst, G. (2013). Potential for carfree development in the UK. *Urban Design and Planning, 166*(DP2), 136–145.

Mokhtarian, P., & Cao, X. (2008). Examining the impacts of residential self-selection on travel behaviour: A focus on methodologies. *Transportation Research Part B, 42*, 204–228.

Naess, P. (2009). Residential self-selection and appropriate control variables in land use: Travel studies. *Transport Reviews, 29*(3), 293–324.

Roads & Maritime Services. (2013). *Guide to traffic generating developments: Updated traffic surveys*. Australia: New South Wales Government.

Roads and Traffic Authority. (2002). *Guide to traffic generating developments*. Australia: New South Wales.

Rye, T. (2002). Travel plans: Do they work? *Transport Policy, 9*, 287–298.

WSP. (2014). *Does car ownership increase car use? A study of the use of car parking within residential developments in London*, Commissioned by the Berkeley Group, UK.

Howlett, R., & Watson, T. (2010). Travel planning in Victoria—a new strategic approach to sustaining communities. in *Paper Presented to 33rd Australasian Transport Research Forum (ATRF)*, Canberra, Australia.

Chapter 9
Opportunities to Enhance Impacts

9.1 Introduction

This chapter, as positioned in Fig. 9.1, takes the findings from Chaps. 5–8 and views these through the lens of both implementation theory and planning enforcement theory (introduced earlier in Chap. 3). An integrated theory of implementation and enforcement is then developed, corresponding to research component 5. Table 9.1 details the research gap, opportunity and objective associated with this research component.

In line with research objective 4, the aim of this chapter is to identify and assess opportunities for enhancing the implementation (and subsequent impacts) of travel plans for new residential developments. This is achieved through the application and integration of implementation and planning enforcement theories.

This chapter is structured as follows. Implementation theory is first applied to the research findings using both top-down and bottom-up approaches to implementation. Planning enforcement theory is then applied with consideration given to both systematic and facilitative approaches. An integrated theory of implementation and enforcement is then presented to guide future travel planning practice for new residential developments. As part of this, the extent to which the research findings support the integrated theory is assessed. The chapter concludes with a summary of the opportunities for enhancing the impacts of travel plans for new residential developments.

C. De Gruyter, *Travel Plans for New Residential Developments: Insights from Theory and Practice*, Springer Theses,
DOI 10.1007/978-981-10-2092-6_9

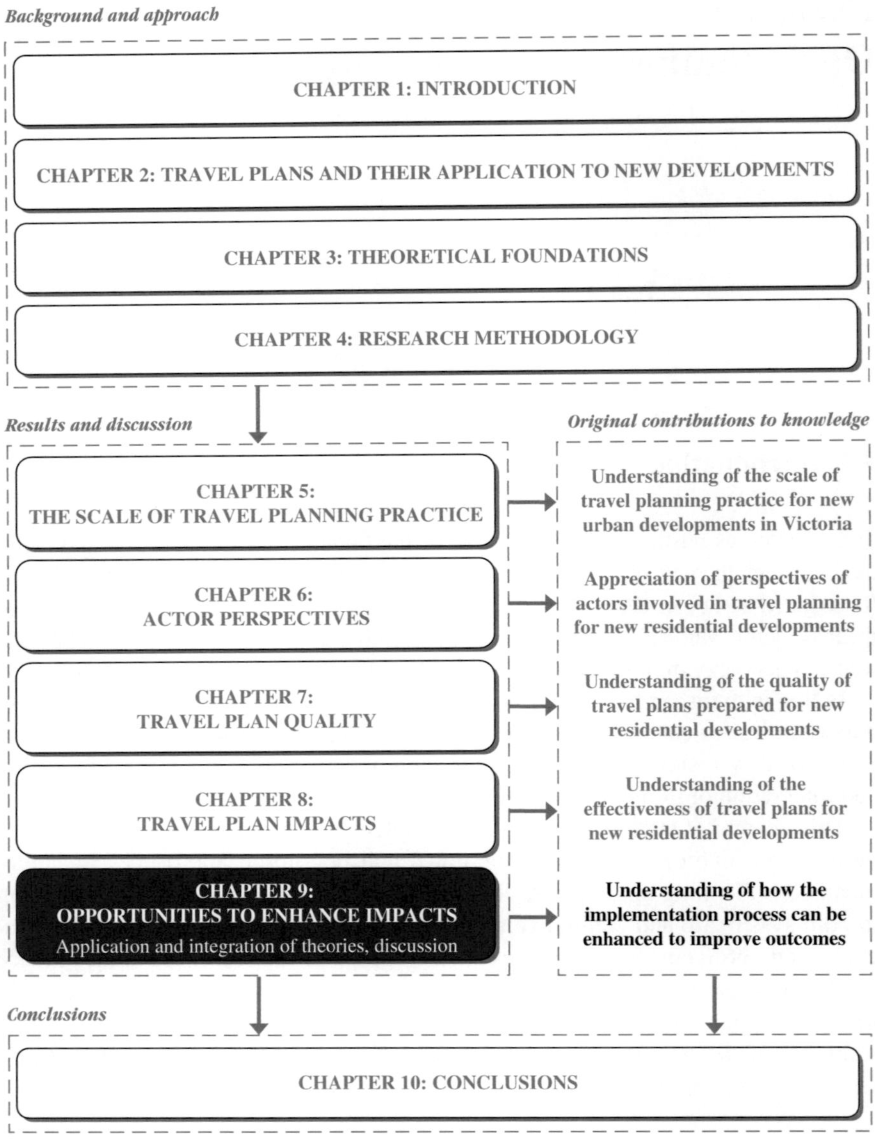

Fig. 9.1 Position of Chap. 9 in the thesis structure

Table 9.1 Research gap, opportunity and objective associated with research component 5

Research gap →	Research opportunity →	Research objective →	Research component
No research has been undertaken to sufficiently explore implementation in the context of travel plans for new residential developments	Explore the implementation process associated with travel plans for new residential developments to identify opportunities to enhance effectiveness	4. To identify and assess opportunities for enhancing their implementation	5. Application and integration of implementation and planning enforcement theories

9.2 Application of Implementation Theory

Implementation theory was introduced in Chap. 3, covering both the top-down and bottom-up approach. This section applies this theory to the research findings. Based on this, opportunities for enhancing the impact of travel plans for new residential developments are identified and discussed.

9.2.1 Application of the Top-down Approach to Implementation

Table 9.2 details the researcher's assessment of the extent to which each of the six top-down conditions developed by Sabatier and Mazmanian (1981) for effective implementation are met in the context of travel plans for new residential developments in Australia. The assessment was undertaken based on the research findings, mostly from Chaps. 5 and 6. The conditions developed by Sabatier and Mazmanian (1981) were used in preference to other top-down approaches (e.g. Gunn 1978) given their extensive exposure to empirical testing in the public policy field (Parsons 1995; Sabatier 1986).

As shown in Table 9.2, conditions 1 and 5 are mostly satisfied, as residential travel planning objectives are generally well established and government agencies are relatively supportive of the concept (as evidenced in Chaps. 6 and 7). However, significant gaps exist in meeting conditions 2 and 3, primarily due to insufficient monitoring and enforcement of travel plans, combined with the lack of any robust planning or legal requirement (as reported in Chaps. 5 and 6).

Figure 9.2 illustrates the same set of information but also provides a comparison to England. This assessment draws upon the responses from English representatives who were interviewed, (as detailed in Chap. 6), but is also based upon relevant literature specific to England (Addison & Associates 2008; Department for Transport 2005; Enoch 2012; Enoch and Ison 2013; Morris et al. 2009; Rye et al. 2011a). As shown by Fig. 9.2, the conditions for effective implementation are generally met to a greater extent in England where a National Planning Policy Framework is in place that is supportive of travel plans (Department for Communities and Local Government 2012) and the use of legal agreements (known as section 106 agreements) are more commonly used to secure travel plans through the planning process (Rye et al. 2011a). In addition, England has a longer history of residential travel planning and therefore presumably a greater level of experience with implementation. One condition under which England does not perform as well as Australia is condition 6 (changes in socio-economic conditions that do not undermine political support or causal theory). This difference is attributed to the UK government scaling back their support for travel plans following the election of a more conservative government in 2010 (Enoch and Ison 2013).

Table 9.2 Extent to which top-down conditions for effective implementation are met in the context of travel plans for new residential developments in Australia

Top-down conditions of effective implementation[a]	Extent to which condition is met[b]	Discussion
1. Clear and consistent objectives	+	Objectives relating to reduced car use are generally well established. However, the objectives of some actors may differ according to their specific goals, e.g. some developers may simply wish to seek planning approval
2. Adequate causal theory (establishment of the link between the problem and solution)	−	Insufficient monitoring and enforcement of residential travel plans has led to a lack of evidence regarding their effectiveness. However, the link between increasing car traffic brought about by new residential developments and the measures available to offset this are well recognised
3. Implementation process legally structured to enhance compliance	- -	Limited resources within local government to enforce residential travel plans, combined with the lack of any robust planning or legal requirement has adversely affected implementation and therefore compliance
4. Committed and skilful implementing officials	±	Some actors involved in residential travel planning have experience with implementation, although most are only involved in requiring or preparing residential travel plans
5. Support of interest groups and sovereigns	+	Industry representatives are generally supportive of travel plans for new residential developments although they have limited confidence that they can be implemented successfully. Local government appear to be more supportive than other types of organisations
6. Changes in socio-economic conditions that do not undermine political support or causal theory	±	Requirements for residential travel plans may be vulnerable to changes in future government transport policy, yet they have managed to sustain a number of government policy changes to date

[a]Based on Sabatier and Mazmanian (1981)
[b]Based on researcher's assessment; interpretation of ratings as follows
++ Very High
+ High
± Moderate
- Low
- - Very Low

While Fig. 9.2 shows that the conditions for effective implementation are generally met to a greater extent in England, the literature suggests that travel plan implementation still remains an issue in England (Rye et al. 2011a). This was also confirmed by the English representatives who were interviewed. This gap between

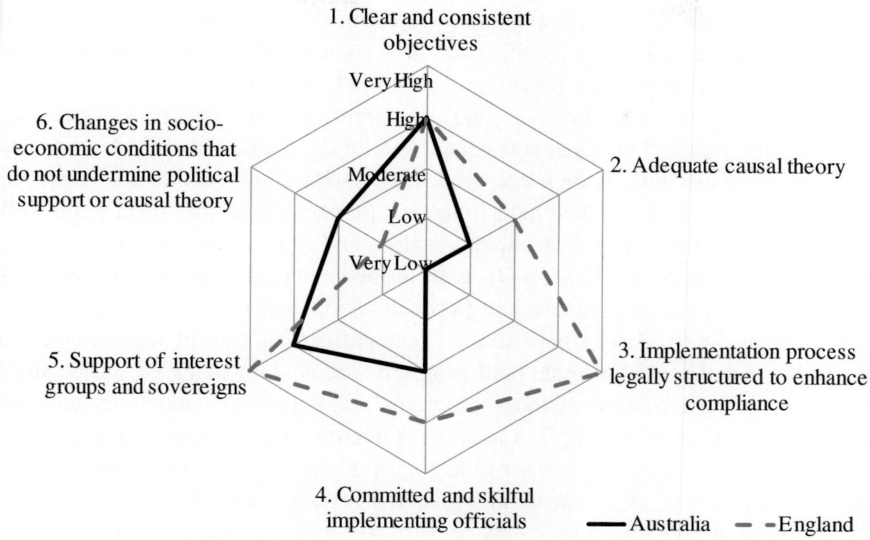

Fig. 9.2 Extent to which top-down conditions for effective implementation are met in the context of travel plans for new residential developments in Australia and England

theory and practice presents an opportunity to identify other conditions that may be important for the effective implementation of travel plans for new residential developments. Firstly, the role of enforcement, while related to condition 3 (implementation process legally structured to enhance compliance), could be made more explicit given that travel plans for new residential developments typically arise out of a planning requirement. Secondly, while condition 4 considers the commitment and skill of implementing officials, the role that dedicated funding plays in supporting the provision of adequate resources for implementation needs to be acknowledged. These additional 'conditions' are considered later in this chapter in developing an integrated theory of implementation and enforcement. However, the roles and preferences of 'implementers' also need to be taken into account given their ability to affect the implementation process. These are considered next through the application of the bottom-up approach.

9.2.2 Application of the Bottom-up Approach to Implementation

O'Toole (2007, p. 147) suggests that the number of actors involved in delivering a given policy can affect the probability of implementation success and that with 'sequential arrangements, adding more organizational units in a chain increases the number of possible roadblocks to action'. In applying this theory, Fig. 9.3 depicts the key actors involved in travel planning for new residential developments, derived from Chap. 6, as a series of links in a chain. The key phases involved in the delivery of residential travel planning are also shown with implementation represented as just one step in the process, albeit a vital one. As can be seen, the target group that comprises residents is a long way from the initial government decision to require the travel plan. Each subsequent link in the chain is therefore crucial to ensuring the travel plan is then prepared, implemented and monitored. However, there is often a 'break' in the chain following preparation of the travel plan. This is supported by the finding that implementation and monitoring has received little attention (evidenced by Chaps. 5–7), with property managers having had little or no involvement in the process thus far (as detailed in Chap. 6).

Elmore (1978, p. 209) argues that a 'frequent explanation of implementation failures is that those who implement programs are seldom included in decisions that determine the content of those programs'. This has particular relevance for the implementation of travel plans at new residential developments in that government agencies are typically responsible for framing the travel plan requirement, with developers and their consultants involved in identifying the measures to be implemented (as evidenced by Chap. 6). Property and building managers are then left with the travel plan to implement, despite them having no prior involvement in its development.

9.2.3 Opportunities to Enhance the Impacts of Travel Plans for New Residential Developments

Based on the application of implementation theory, a number of opportunities have been identified to enhance the impacts of travel plans for new residential developments. These include:

- Ensuring planning requirements are sound and supported by relevant planning policies
- Developing a stronger industry focus for residential travel planning
- Facilitating greater ownership and engagement in the residential travel planning process
- Improving the quality of residential travel plans
- Developing guidance material that is tailored to new residential developments
- Providing regular training opportunities and forums for sharing knowledge.

Each of these opportunities is now discussed in turn.

9.2.3.1 Ensuring Planning Requirements are Sound and Supported by Relevant Planning Policies

Planning requirements for residential travel plans need to be clearly specified to ensure they are interpreted correctly, instigate an appropriate level of participation

Fig. 9.3 Key actors in the travel planning process for new residential developments

in the process, and facilitate an intended outcome. They also need to be applied consistently to ensure the process is equitable for property developers. However, as identified in Chap. 6, planning requirements for residential travel plans also need to be sufficiently flexible so that location specific circumstances can be taken into account.

In line with the top-down approach and the research findings from Chaps. 5 and 6, planning requirements for residential travel plans also need to be supported by relevant planning policies. The lack of any state or national planning policy in Australia that is supportive of travel plans contrasts that of the United Kingdom, Sweden and Switzerland, where relevant planning policies are in place at the national level (Rye et al. 2011b). As outlined in Chap. 2, supportive planning policy is recognised as a key success factor in requiring travel plans for new residential developments (Addison & Associates 2008; Department for Transport 2005).

9.2.3.2 Developing a Stronger Industry Focus for Residential Travel Planning

Developing a stronger industry focus for residential travel planning would help to deal with the limited amount of experience in implementation, as identified in Chap. 6. Independent third parties, for example as not-for-profit associations, could be established to support the implementation and monitoring of residential travel plans. These third parties could be funded by developers to include regular monitoring reports to local government on progress and outcomes. This would also provide the opportunity to ensure that implementation is tailored to reflect the characteristics of residential developments and their associated management structures.

9.2.3.3 Facilitating Greater Ownership and Engagement in the Residential Travel Planning Process

In line with the bottom-up approach to implementation, greater ownership and engagement could be facilitated through the involvement of 'implementers', such as property and building managers, earlier in the travel planning process. Where possible, those responsible for implementation should be involved in developing the travel plan so that they have a stronger sense of ownership when implementing the travel plan. This approach is consistent with the travel planning success factors outlined in Chap. 2 and also reflects the suggestions made by interview participants reported in Chap. 6.

9.2.3.4 Improving the Quality of Residential Travel Plans

To increase the likelihood of successful implementation, the quality of residential travel plans can be improved. Chapter 7 showed that greater attention should be directed towards estimating the expected travel patterns of residents, specifying how the travel plan will be managed and implemented (including key responsibilities and funding), and outlining clear processes for monitoring and review. To facilitate these improvements, councils and other relevant authorities could seek greater involvement from experienced practitioners (sourced in-house or externally) to review and subsequently improve the quality of residential travel plans prior to granting planning approval. These reviews would need to be cognisant of the differences associated with residential travel plans such as the presence of different management structures and the need to cater for a range of trip purposes and destinations. Involvement from property and building managers at this stage may also help to ensure that the measures proposed within the travel plans are both reasonable and appropriate.

Where possible, the process used to assess the quality of residential travel plans should be made transparent to those involved in preparing the travel plans, such as property developers. This can also be made possible through the development of guidance material.

9.2.3.5 Developing Guidance Material That is Tailored to New Residential Developments

Despite the excellent range of travel planning guidance available, particularly in the United Kingdom (Department for Transport 2009; Transport for London 2011b), limited information is available on applying travel planning principles to residential sites (Morris et al. 2009). This is of particular note given the differences associated with residential travel plans compared to the more traditional workplace and school travel plans.

There is only one guideline currently available on travel planning specifically for new residential developments (Department for Transport 2005), although is mostly tailored to the United Kingdom. Guidance material therefore needs to be developed and tailored to new residential developments for other jurisdictions, including Australia. This was raised in the interviews reported in Chap. 6, along with the need to provide sufficient guidance on implementation responsibilities, including funding of the travel plan. Guidance material should also highlight the need to actively promote travel plans beyond development occupation, given the turnover of new residents over time. Ongoing promotion is also important given the findings from Chap. 8 which showed relatively low awareness of some travel plan initiatives at the case sites.

The need for guidance material is further evidenced by the findings of the council survey (Chap. 5) which revealed a relatively low level of practical experience in travel planning among council representatives. Guidance material should therefore support councils in requiring travel plans for new residential developments, including methods for assessing their quality.

9.2.3.6 Providing Regular Training Opportunities and Forums for Sharing Knowledge

The provision of regular training opportunities to build capacity of the industry is critical to the successful delivery of residential travel plans. This is particularly important for property managers, should they become more involved in the implementation of residential travel plans. Training opportunities for council representatives would also be beneficial given their limited practical experience in travel planning, as identified in Chap. 5. As part of any capacity building effort, forums delivered through a range of mediums will be crucial for sharing knowledge and experiences among industry practitioners. Where possible, training programs should also incorporate the use of relevant travel planning guidance (Rye et al. 2011a).

9.3 Application of Planning Enforcement Theory

Planning enforcement theory was introduced earlier in Chap. 3, covering both the systematic and facilitative approach. A conceptual representation of these approaches using the planning enforcement pyramid is presented again in Fig. 9.4.

This section applies planning enforcement theory to the research findings. Based on this, opportunities for enhancing the impact of travel plans for new residential developments are identified and discussed. However, before doing so, some additional context is provided in terms of the options available for requiring and enforcing travel plans for new residential developments through the land use planning and approvals process.

9.3.1 Context

Travel plans are generally required for new developments (including residential sites) by way of a planning condition or formal agreement. Basic principles have been established around the validity of planning conditions. Each condition must be reasonable and relevant, fulfil a planning purpose, accurately convey its intended effect, and avoid uncertainty and vagueness. In addition, planning

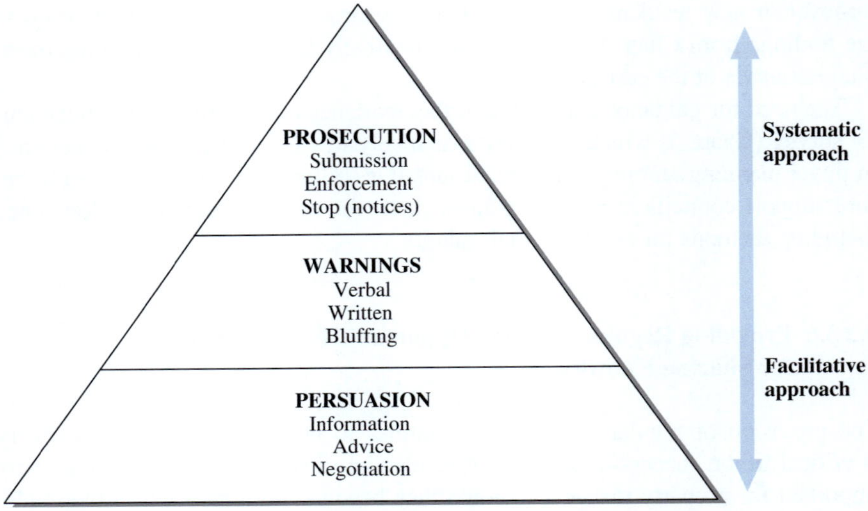

Fig. 9.4 Planning enforcement pyramid. *Source* Author's adaptation based on McKay (2003)

conditions must be enforceable (Department of Transport, Planning and Local Infrastructure 2014). Formal agreements generally arise out of planning legislation and can be registered over the title of the land and become binding upon future owners (Department of Transport, Planning and Local Infrastructure 2014).

Options available to local government for enforcing planning conditions and formal agreements range from negotiation and official warnings, to enforcement orders and court proceedings (Department of Transport, Planning and Local Infrastructure 2014), thereby encompassing all layers of the planning enforcement pyramid shown in Fig. 9.4. While planning conditions are generally intended to be met prior to occupation, they can be enforced at any time beyond occupation of the development (Planning Enforcement Officers Association Inc. 2007).

9.3.2 Application of the Systematic Approach to Planning Enforcement

The systematic approach to planning enforcement is concerned with the uniform and strict application of rules. Legislative mechanisms, such as sanctions and fines, are typically used to deter violations (Prior 2000). As outlined by Prior (2000), systematic enforcement assumes that:

- Breaches of regulations are essentially intended
- Most perpetrators are aware of required rules and standards

- The threat of punitive sanctions is an essential deterrent to potential violation
- A comprehensive approach to enforcement is essential
- Rules are clear and unambiguous
- Regulators are effectively resourced and empowered
- Enforcement actions derive from reactions to violations.

Given the relative lack of enforcement of travel plans for new residential developments to date, it is difficult to determine which of the above assumptions apply in practice. However, there is sufficient evidence to suggest that a number of the assumptions are unlikely to be valid. For example, the general lack of guidance material means that 'rules' are *not* necessarily 'clear and unambiguous', as highlighted by a property developer who was interviewed (see Chap. 6):

> There's no rules about this, it's very unregulated. It's not clear in terms of what you have to do or why you have to do it [PD1].

In addition, the results of the council survey reported in Chap. 5 suggest that local government, as regulators, are *not* 'effectively resourced and empowered'. One of the key reasons reported by Victorian councils for the relatively low rate of monitoring was a lack of resources. This finding is also consistent with similar surveys of local authorities undertaken in the United Kingdom (Addison & Associates 2008; Llewellyn et al. 2014). The systematic approach also assumes that 'enforcement actions derive from reactions to violations'. However, it may be questionable as to whether a complaint would arise from inaction of a travel plan, particularly if awareness of travel plan initiatives is low (as was identified in Chap. 8) or if there are no perceived traffic and parking issues at the development.

Despite evidence to suggest that the systematic approach may not be entirely appropriate for enforcing travel plans for new residential developments, the theory on planning enforcement recommends its use as a last resort when all other options are exhausted. This is considered necessary to protect the integrity of the planning system, particularly in dealing with instances of repeat and flagrant offenders (Burby et al. 1998; Harris 2010; McKay 2003).

Given the limited applicability of systematic enforcement to residential travel plans, it is appropriate to consider the applicability of the facilitative approach to planning enforcement.

9.3.3 Application of the Facilitative Approach to Planning Enforcement

The facilitative approach to planning enforcement favours the use of incentives, negotiation and education to assist offenders to comply with regulations (Burby et al. 1998; McKay 2003). It is based on the assumption that most breaches of regulations occur through ignorance and are therefore unintended (McKay 2003; Prior 2000). The facilitative approach is considered to be well suited to situations where resources for enforcement are limited as the approach is less resource intensive than systematic enforcement regimes (Harris 2011).

Given that travel plans are a relatively new concept for the property development industry, with limited training and guidance material currently available, an educational style of enforcement consistent with the facilitative approach would appear to be appropriate. This method is also consistent with guidance available on planning enforcement which emphasises the need to obtain compliance over prosecuting offenders (Planning Enforcement Officers Association Inc. 2007). The less resource intensive nature of facilitative enforcement would also appear favourable given the lack of council resources currently available for planning enforcement, as identified in Chaps. 5 and 6.

9.3.4 Opportunities to Enhance the Impacts of Travel Plans for New Residential Developments

Based on the application of planning enforcement theory, a number of opportunities have been identified to enhance the impacts of travel plans for new residential developments. These include:

- Adopting a more pro-active and facilitative style of enforcement but retaining the option to employ a systematic approach if needed
- Ensuring an adequate number of technically competent staff are available for enforcement
- Incorporating best practice elements of enforcement into training and guidance material
- Extending training opportunities to enforcement officers.

Each of these opportunities is now discussed in turn.

9.3.4.1 Adopting a More Pro-active and Facilitative Style of Enforcement but Retaining the Option to Employ a Systematic Approach if Needed

A facilitative approach should be adopted to enforce travel plans for new residential developments, in line with the theory and guidance on planning enforcement. Education and advice should form key elements of this approach given that travel plans are a relatively new concept for the property development industry. This would also help to ensure that travel planning is viewed in a positive light and not necessarily seen as a burden. However, the option to employ systematic enforcement methods, such as sanctions and fines, should still be retained to deal with any repeat and flagrant offenders, and to protect the integrity of the planning system.

Planning enforcement is generally undertaken in response to public complaints (Planning Enforcement Officers Association Inc. 2007). However, as a lack of travel plan implementation may not necessarily result in public complaints, a shift from a reactive to pro-active enforcement culture will be required. In doing so, this approach will ultimately help to boost implementation rates associated with travel plans for new residential developments.

9.3.4.2 Ensuring an Adequate Number of Technically Competent Staff are Available for Enforcement

While the facilitative approach tends to require less resources than systematic enforcement regimes, there is still a need to ensure an adequate number of technically competent staff are available for enforcement (Burby et al. 1998). However, the responsibility for enforcing travel plans at new residential developments should not lie solely with enforcement officers in councils. Other council representatives involved in requiring travel plans, such as planners and traffic engineers, can also be involved in enforcement, particularly given that an emphasis should be placed on providing education and advice on travel planning matters.

9.3.4.3 Incorporating Best Practice Elements of Enforcement into Training and Guidance Material

Training and guidance material on travel plans for new residential developments should reflect best practice elements of enforcement, particularly features of the facilitative approach and how they can be applied to the enforcement of travel plans. Including this information in training and guidance material will become particularly relevant should other council representatives become more involved in enforcement.

9.3.4.4 Extending Training Opportunities to Enforcement Officers

Training opportunities relating to travel planning should extend to enforcement officers to ensure they are familiar with travel plans and the types of enforcement styles that are appropriate. Moreover, training should include a component on enforcement, with an emphasis on building necessary skills in verbal and written communication, negotiation, and conflict resolution (Victorian Auditor-General 2008).

9.3.4.5 Other Opportunities

While not directly related to planning enforcement, the research findings from Chap. 8 suggest two additional opportunities for enhancing travel plans for new residential developments, albeit in the context of monitoring. Firstly, given that data comparability issues may be experienced when using secondary data sources, control sites should be used where possible to provide a more accurate indication of travel plan effectiveness. Secondly, the potential for residential self-selection suggests that, where possible, this phenomenon should be controlled for in future evaluations to avoid overstating the impacts of travel plans.

9.4 Towards an Integrated Theory of Implementation and Enforcement

This section presents the development of an integrated theory of implementation and enforcement to guide future travel planning practice for new residential developments. This includes an assessment of the extent to which the research findings support the integrated theory.

Figure 9.5 provides a conceptual representation of existing implementation theory and planning enforcement theory, as presented earlier in Chap. 3. Key characteristics of the top-down and bottom-up approaches to implementation are shown, along with key features of the systematic and facilitative approaches to enforcement. The dashed lines dividing the approaches denote their consideration as distinct and separate entities.

Rather than solely taking a top-down or bottom-up approach to implementation, the application of implementation theory to the research findings has shown that both approaches have merit in the context of travel plans for new residential developments. The top-down approach provided a clear set of conditions for assessing implementation but was not able to sufficiently account for the roles and preferences of various actors involved in the process, as was possible with the bottom-up approach. It is therefore appropriate that the top-down and bottom-up approaches to implementation are combined when studying the implementation of travel plans for new residential developments. This approach is consistent with the findings of Pülzl and Treib (2007) who note that there is now general agreement that implementation is located on a continuum between central authority (top-down) and local autonomy (bottom-up). They suggest that the 'preferences of street-level bureaucrats and the negotiations within implementation networks have to be taken into account to the same extent as centrally defined policy objectives and efforts at hierarchical control' (Pülzl and Treib 2007, p. 100).

In a similar manner, rather than solely taking a systematic or facilitative approach to enforcement, the application of planning enforcement theory to the research findings has shown that both approaches have a role to play. While the facilitative approach was found to be more appropriate for enforcing travel plans for new residential developments, the option to employ systematic methods was also seen as important, albeit as a last resort to deal with the possibility of repeat and flagrant offenders, and to protect the integrity of the planning system. This combined approach, with a skew towards facilitative enforcement, aligns with the

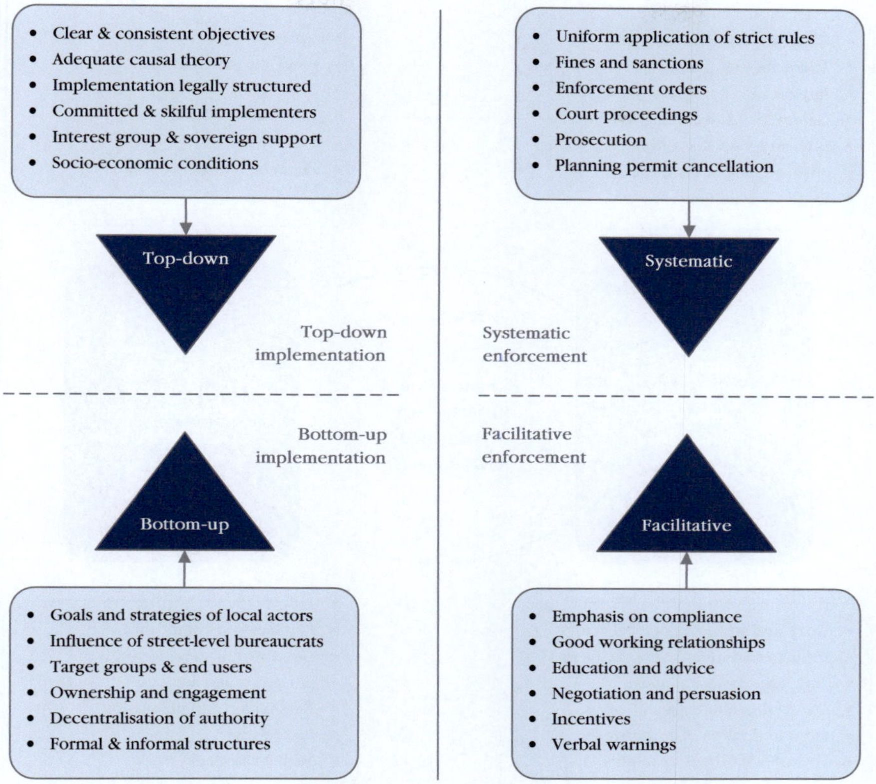

Fig. 9.5 Conceptualisation of implementation theory and planning enforcement theory

theory and guidance on planning enforcement (Burby et al. 1998; Harris 2010; McKay 2003; Planning Enforcement Officers Association Inc. 2007).

In applying top-down implementation theory to the research findings, three additional conditions for effective implementation were identified, reflecting a gap between existing theory and practice. The first condition related to the role of enforcement and how this could be made more explicit given that travel plans for new residential developments typically arise out of a planning requirement. By integrating implementation theory with planning enforcement theory, this condition can be considered as an explicit, yet integrated element of the travel planning process. The second condition related to the role that dedicated funding plays in supporting the provision of adequate resources for implementing travel plans at new residential developments. This condition can be added as a 'seventh' top-down condition of effective implementation, given that funding is usually allocated in a top-down manner. The third and last additional condition related to the roles and preferences of implementers in the travel planning process. By combining the top-down and bottom-up approaches to implementation, this condition can be incorporated.

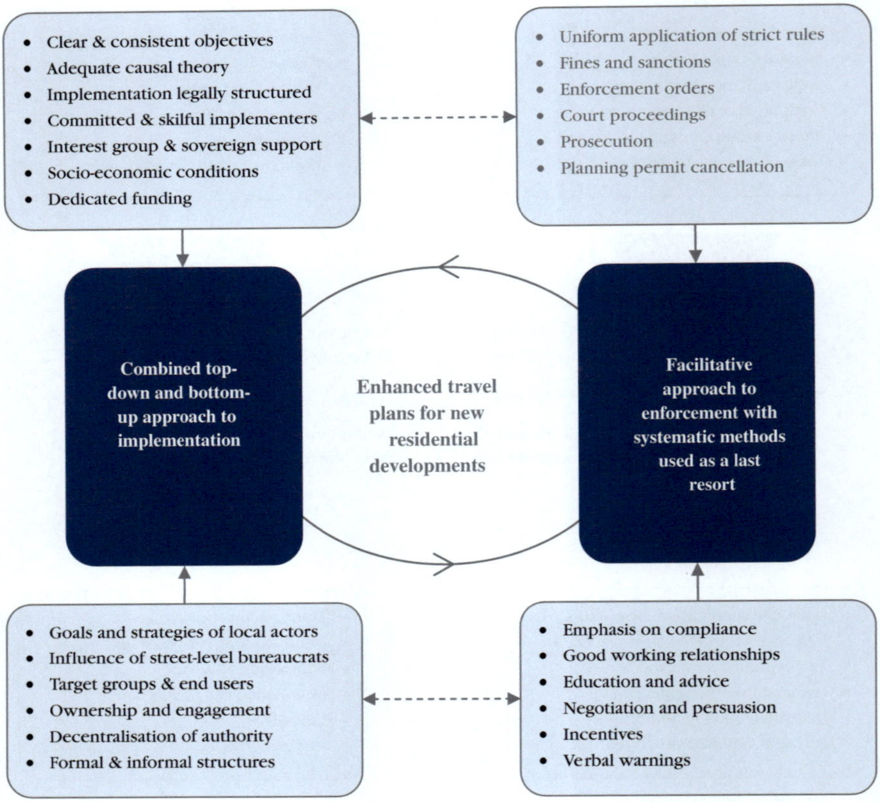

Fig. 9.6 Conceptualisation of integrated theory of implementation and enforcement

Based on these changes, an integrated theory of implementation and enforcement is presented conceptually in Fig. 9.6, with circular arrows used to denote the integration of these elements. Figure 9.6 also shows how the top-down and bottom-up approach to implementation is combined, with the inclusion of 'dedicated funding' as an additional top-down condition. The facilitative approach is also combined with the systematic approach, albeit with a stronger focus on the facilitative style of enforcement. Here, features of the systematic approach have been 'greyed out' to reduce their emphasis, in line with the desire to employ these methods only as a last resort. In addition, the similarities between top-down implementation and systematic enforcement, and bottom-up implementation and facilitative enforcement, are indicated by dashed arrows, noting that differences are still inherent in their fundamental purpose.

By integrating implementation and enforcement, travel planning for new residential developments can be enhanced through improvements to both the quality and consistency of implementation. As highlighted by this discussion, the research findings have shown support for an integrated theory of implementation and enforcement. The need to combine top-down and bottom-up approaches to implementation has been evidenced, as has the importance of retaining systematic methods while primarily adopting a facilitative approach to enforcement. Furthermore, the need to consider implementation and enforcement as an integrated approach to travel planning for new residential developments has also been supported by the research findings.

9.5 Conclusion

The aim of this chapter is to identify and assess opportunities for enhancing the implementation (and subsequent impacts) of travel plans for new residential developments. In doing so, implementation theory and planning enforcement theory were applied to the research findings. An integrated theory of implementation and enforcement was then developed to guide future travel planning practice for new residential developments.

In applying implementation theory to the research findings, a number of opportunities were identified for enhancing the impacts of travel plans for new residential developments. In the short term, these opportunities include: facilitating greater ownership and engagement in the process through earlier involvement of 'implementers', improving the quality of residential travel plans prior to granting planning approval, developing tailored guidance material, and providing regular training opportunities and forums for sharing knowledge among practitioners. In the longer term, the development of sound planning requirements, with sufficient flexibility to take into account local circumstances, will help to provide greater clarity in the requirements for travel plans for new residential developments. This will need to be backed by the development of a stronger industry focus for residential travel planning, with recognition of the diverse set of actors currently involved in the process.

Application of planning enforcement theory to the research findings has also identified opportunities for enhancing the impacts of travel plans for new residential developments. In the short term these include: adopting a more pro-active and facilitative style of enforcement yet retaining the option to employ a systematic approach if needed, incorporating best practice elements of enforcement into training and guidance material, and extending training opportunities to enforcement officers. In the longer term, there is a need to ensure an adequate number of technically competent staff are available for enforcement, including the involvement of other council representatives where appropriate.

A summary of the opportunities for enhancing the impacts of travel plans for new residential developments is presented in Table 9.3. Acting on these opportunities is imperative for enhancing travel planning at new residential developments. This need is further supported by the survey findings reported in Chap. 5 in which half of the councils indicated they were likely to continue to require travel plans. It is also supported by the interview findings reported in Chap. 6 in which the

Table 9.3 Summary of opportunities for enhancing impacts

Enhancement area	Opportunities and potential actions	Timeframe
Ownership and engagement	• Facilitate greater ownership and engagement in the travel planning process through earlier involvement of 'implementers'	Short term
Travel plan quality	• Improve the quality of residential travel plans by assessing them against a best practice framework prior to granting planning approval • Seek greater involvement from experienced practitioners to review and subsequently improve travel plan quality • Ensure the assessment process is made transparent to all parties	Short term
Guidance and training	• Develop guidance material that is tailored to new residential developments and incorporates best practice in enforcement • Provide regular training opportunities for sharing knowledge among industry practitioners, including enforcement officers	Short term
Enforcement styles	• Adopt a more pro-active and facilitative style of enforcement but retain the option to employ a systematic approach if needed	Short term
Evaluation	• Use control sites where possible to provide a more accurate indication of travel plan effectiveness • Control for residential self-selection effects where possible to avoid overstating the impacts of residential travel plans	Short term
Planning requirements	• Ensure planning requirements are sound and supported by relevant planning policies, with sufficient flexibility so that location specific circumstances can be taken into account	Long term
Industry focus and capacity	• Develop a stronger industry focus for residential travel planning, potentially through the involvement of not-for-profit associations • Ensure an adequate number of technically competent staff are available for enforcement, including other council representatives	Long term

majority of industry representatives felt that travel plans for new residential developments are either here to stay or will increase in focus.

This chapter has discussed the application and integration of implementation and planning enforcement theory to guide future travel planning practice in the context of new residential developments. Consideration of both top-down and bottom-up approaches to implementation, a facilitative style of enforcement with systematic means adopted if needed, and the integration of implementation and enforcement, will help to improve both the quality and consistency of travel plans that are implemented at new residential developments in the future.

The next and final chapter of this thesis presents a set of conclusions to the research. This includes a summary of key contributions and directions for future research.

References

Addison & Associates. (2008). *Delivering travel plans through the planning process—Research report*. London, UK: Department for Transport and Communities and Local Government.

Burby, R., May, P., & Paterson, R. (1998). Improving compliance with regulations: Choices and outcomes for local government. *Journal of the American Planning Association, 64*(3), 324–334.

Department for Communities and Local Government. (2012). *National planning policy framework*, London, UK.

Department for Transport. (2005). *Making residential travel plans work: Guidelines for new development*, London, UK.

Department for Transport. (2009). *Good practice guidelines: Delivering travel plans through the planning process*. London, UK: Department for Transport.

Department of Transport, Planning and Local Infrastructure. 2014. *A guide to the planning system*, viewed 25 September 2014. http://www.dpcd.vic.gov.au/planning/.

De Gruyter, C., Rose, G., & Currie, G. (2014). 'Securing travel plans through the planning approvals process: A case study of practice from Victoria, Australia', *Cities, 41*, part A, 114–122.

De Gruyter, C., Rose, G., & Currie, G. (2015). 'Enhancing the impact of travel plans for new residential developments: Insights from implementation theory', *Transport Policy, 40*, 24–35.

Elmore, R. (1978). Organizational models of social program implementation. *Public Policy, 26*(2), 185–228.

Enoch, M. (2012). *Sustainable transport, mobility management and travel plans*. Surrey, England: Ashgate Publishing Limited.

Enoch, M., & Ison, S. (2013). Travel plans: A way forward? *Proceedings of the Institution of Civil Engineers: Urban Design and Planning, 166*(DP2), 126–135.

Gunn, L. A. (1978). Why is implementation so difficult? *Management Services in Government, 33*, 169–176.

Harris, N. (2010). Discretion and expediency in the enforcement of planning controls. *Town Planning Review, 81*(6), 675–700.

Harris, N. (2011). Discipline surveillance, control: A foucaultian perspective on the enforcement of planning regulations. *Planning Theory & Practice, 12*(1), 57–76.

Llewellyn, R., Paton, D., & Tricker, R. (2014). *After the downturn: Where now for travel plans in Scotland?* Paper presented to Scottish Transport Applications & Research (STAR) conference, Glasgow, Scotland.

Mazmanian, D., & Sabatier, P. (1981). *Effective policy implementation*. US: Lexington Books, D.C. Heath and Company.

McKay, S. (2003). Sheriffs and outlaws: In pursuit of effective enforcement. *Town Planning Review, 74*(4), 423–443.

Morris, D., Enoch, M., Pitfield, D., & Ison, S. (2009). Car-free development through UK community travel plans. *Urban Design and Planning, 162*, 19–27.

O'Toole, L. (2007). Interorganizational relations in implementation. In B. Peters & J. Pierre (Eds.), *Handbook of public administration* (pp. 142–152). London, UK: SAGE Publications Ltd.

Parsons, W. (1995). *Public policy: An introduction to the theory and practice of policy analysis*. Glos, UK: Edward Elgar Publishing.

Planning Enforcement Officers Association Inc. (2007). *A guide to planning enforcement in Victoria*, PEOA, Victoria, Australia, viewed 25 September 2014. http://planning-enforcement.com/.

Prior, A. (2000). Problems in the theory and practice of planning enforcement. *Planning Theory & Practice, 1*(1), 53–69.

Pülzl, H., & Treib, O. (2007). Implementing public policy. In F. Fischer, G. Miller & M. Sidney (Eds.), *Handbook of public policy analysis: Theory, politics and methods.* US: Taylor & Francis Group.

Rye, T., Green, C., Young, E., & Ison, S. (2011a). Using the land-use planning process to secure travel plans: An assessment of progress in England to date. *Journal of Transport Geography, 19*(2), 235–243.

Rye, T., Welsch, J., Plevnik, A., & de Tomassi, R. (2011b). First steps towards cross-national transfer in integrating mobility management and land use planning in the EU and Switzerland. *Transport Policy, 18*, 533–543.

Sabatier, P. (1986). Top-down and bottom-up approaches to implementation research: A critical analysis and suggested synthesis. *Journal of Public Policy, 6*(1), 21–48.

Transport for London. (2011). *Travel planning for new development in London—incorporating deliveries and servicing*. UK: Transport for London.

Victorian Auditor-General. (2008). *Enforcement of planning permits.* Australia: Victoria.

Chapter 10
Conclusions

10.1 Introduction

This thesis has explored the use of travel plans for new residential developments. The research undertaken has provided a number of original contributions to knowledge in this field, as presented in previous chapters.

This chapter, as positioned in Fig. 10.1, concludes the thesis by providing a summary of key findings and contributions to demonstrate how the research aim and objectives have been met. Implications for theory and practice are also discussed. A critique of the research approach is then presented, followed by a discussion of future research directions.

10.2 Summary of Key Findings and Contributions

As described in Chap. 1, the aim of this research was:

To assess the effectiveness of travel plans for new residential developments and to identify opportunities to enhance their effectiveness

A number of research objectives were also identified in Chap. 1 as key steps required to meet the research aim. In the context of travel plans for new residential developments, these were:

1. To examine the scale of practice in Victoria, Australia
2. To gain an appreciation for the perspectives of industry actors involved in their application
3. To evaluate their quality and effectiveness
4. To identify and assess opportunities for enhancing their implementation.

Table 10.1 provides a summary of key findings and contributions of this thesis, including their alignment to each of the research objectives.

C. De Gruyter, *Travel Plans for New Residential Developments: Insights from Theory and Practice*, Springer Theses,
DOI 10.1007/978-981-10-2092-6_10

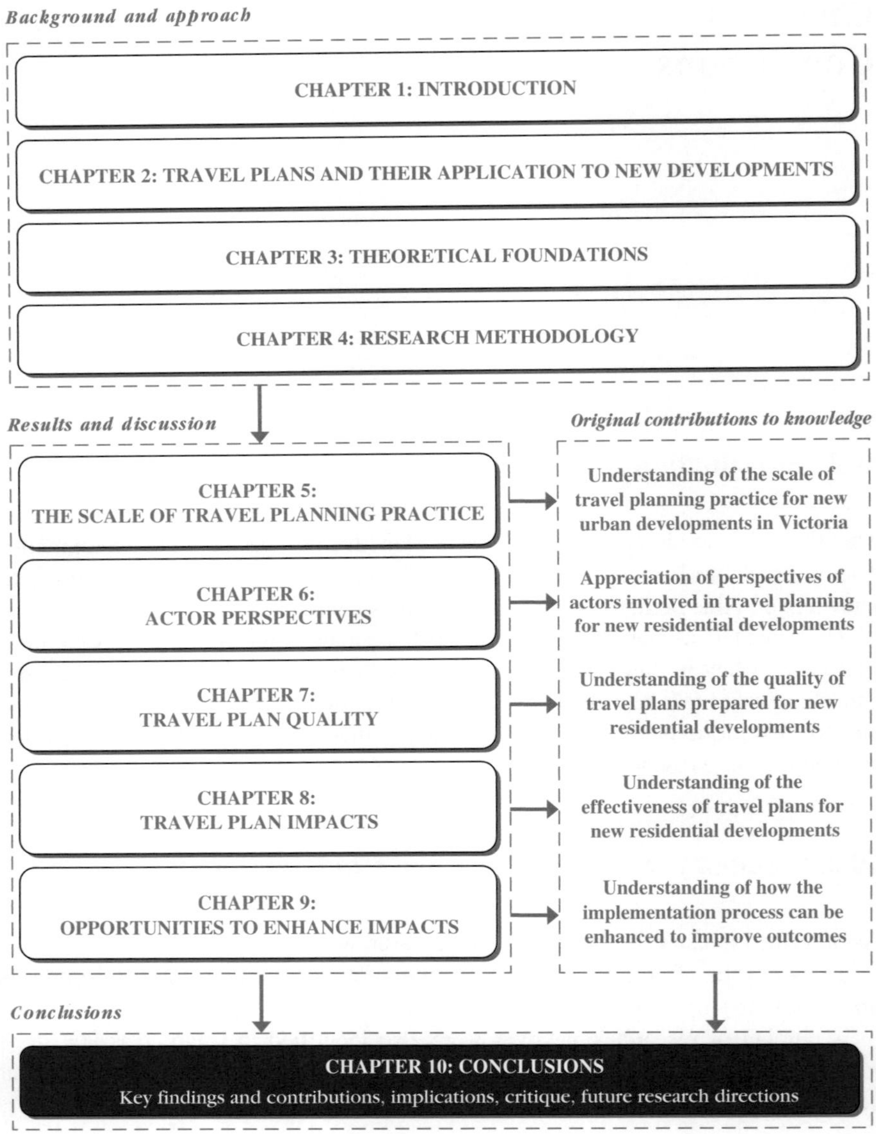

Fig. 10.1 Position of Chap. 10 in the thesis structure

An understanding of the scale of travel planning practice for new urban developments in Victoria was provided in Chap. 5, corresponding to research objective 1. This showed that half of the surveyed councils in Victoria had previously required a travel plan for a new development, with around 100 travel plans required during 2010–12 alone.

Table 10.1 Summary of key findings and contributions

Research objective[a]	Original contribution to knowledge	Key findings
1. To examine the scale of practice in Victoria, Australia	An understanding of the scale of travel planning practice for new urban developments in Victoria, Australia (Chap. 5)	• Half of the Victorian councils surveyed (50 %) had previously required a travel plan for a new development, primarily to offset the impact of providing less car parking • Around 100 travel plans had been required during 2010–12, mostly in inner and middle metropolitan areas • Around 80 % of councils had not monitored any of the travel plans they had required • Around half (51 %) of the councils were likely to require a travel plan for a new development in the future
2. To gain an appreciation for the perspectives of industry actors involved in their application	An appreciation for the perspectives of actors involved in travel planning for new residential developments (Chap. 6)	• The industry has had little involvement with implementing and monitoring travel plans for new residential developments to date; most involvement has been focused on preparing/developing travel plans • Actors are generally supportive but have limited confidence in the ability to achieve successful implementation • Challenges associated with implementing travel plans for new residential developments include a lack of enforcement, uncertainty about implementation responsibilities, and a lack of ownership • Potential solutions to implementation challenges include developing a more robust planning/legal requirement, encouraging more resident engagement, and having the developer fund the implementation of the travel plan

(continued)

Table 10.1 (continued)

Research objective[a]	Original contribution to knowledge	Key findings
3. To evaluate their quality and effectiveness	An understanding of the quality of travel plans prepared for new residential developments (Chap. 7)	• Travel plans for new residential developments have been prepared predominantly by consultants • On average, only 47 % of the maximum possible score for travel plan quality was achieved, suggesting considerable scope exists for improving the quality of travel plans for new residential developments • Key areas identified to improve travel plan quality include the estimation of expected travel patterns, specifying how the travel plan will be managed and implemented, and outlining clearer processes for monitoring and review
	An understanding of the effectiveness of travel plans in reducing car use at new residential developments (Chap. 8)	• Average (7 am–9 am) weekday mode share for car driver trips was 14 % points lower at new residential developments with travel plans compared to similar residential developments (control sites) without travel plans • Relying solely on comparisons to secondary data sources to assess travel plan effectiveness can lead to an overestimation of travel plan impacts due to inconsistencies in geographical locations and data collection periods • Relatively low use, and in some cases, awareness of travel plan initiatives was reported by residents • Residential self-selection potentially accounted for 10–42 % of the observed difference in travel behaviour

(continued)

Table 10.1 (continued)

Research objective[a]	Original contribution to knowledge	Key findings
4. To identify and assess opportunities for enhancing their implementation	An understanding of how the implementation of travel plans for new residential developments can be enhanced to improve outcomes (Chap. 9)	• Short-term enhancements include greater ownership and engagement of 'implementers', improvements to travel plan quality, provision of guidance material and training, and pro-active and facilitative enforcement • Long-term enhancements include sound planning requirements, a stronger industry focus for residential travel planning, and an adequate number of technically competent staff available for enforcement • An integrated theory of implementation and enforcement, with consideration to both top-down and bottom-up implementation, and facilitative and systematic enforcement, can help to guide future travel planning practice

[a]Research objectives are framed in the context of travel plans for new residential developments

An appreciation for the perspectives of actors involved in travel planning for new residential developments was provided in Chap. 6, corresponding to research objective 2. This showed general support for travel plans at new residential developments but limited confidence in the ability to implement them successfully. It also highlighted a number of challenges and potential solutions associated with implementation.

An assessment of the quality of travel plans prepared for new residential developments was provided in Chap. 7, corresponding to research objective 3. This showed that considerable scope exists to improve travel plan quality, particularly in estimating expected travel patterns of future users, specifying how the travel plan will be managed and implemented, and outlining clearer processes for monitoring and reviewing the travel plan.

An assessment of the effectiveness of travel plans in reducing car use at new residential developments was provided in Chap. 8 (also corresponding to research objective 3). It revealed that car use at new residential developments with travel plans was about 14 % points lower than matched control sites. It also provided some preliminary evidence to suggest that residential self-selection can potentially contribute 10–42 % of the observed difference in travel behaviour associated with travel plans for new residential developments.

An understanding of how the implementation of travel plans for new residential developments can be enhanced was provided in Chap. 9, corresponding to research objective 4. Application of implementation theory and planning enforcement theory helped to identify a number of short and long term opportunities to enhance impacts. An integrated theory of implementation and enforcement was also developed to guide future travel planning practice for new residential developments.

In fulfilling each of the research objectives, the overall research aim has been met. This was achieved through assessing the effectiveness of travel plans for new residential developments (Chap. 8). In addition, based on the findings from Chaps. 5–8, opportunities were identified to enhance effectiveness through the application and integration of implementation and planning enforcement theories (Chap. 9).

In providing these original contributions to knowledge, this research has added to the existing literature concerning travel plans for new residential developments. However, it has also provided an important **theoretical contribution** by integrating implementation theory and planning enforcement theory. Here, the top-down and bottom-up approaches to implementation were combined, with 'dedicated funding' incorporated as an additional top-down condition for effective implementation. Further, the facilitative and systematic approaches to planning enforcement were combined, with an emphasis placed on the facilitative style of enforcement. The link between implementation and planning enforcement was also depicted to illustrate their interrelationship as part of the travel planning process. This theoretical contribution helps to further our understanding of implementation and enforcement in the context of travel plans for new residential developments, but also facilitates a wider understanding in the context of travel plans more generally.

10.3 Implications for Theory and Practice

Given the findings presented in this thesis, it is appropriate to reflect on their implications for theory and practice. In considering the theory first, the research findings suggest that the implementation of travel plans for new residential developments cannot be viewed solely from a top-down or bottom-up perspective. Both approaches need to be considered when planning, administering and evaluating the implementation process. The research findings also highlight the importance of considering planning enforcement theory in the context of travel plans that have been required for new residential developments. Here, a facilitative/educational style of enforcement should be emphasised, with systematic methods retained as a last resort. In addition, the intrinsic link between implementation and enforcement means that the theories underlying these elements should be considered in an integrated manner, rather than separately. Doing so will help to ensure that both the quality and consistency of implementation is improved.

The research findings also have a number of implications for practice. First, sufficient resources will be required from government and property developers to act on the opportunities identified in Chap. 9 for enhancing travel plan effectiveness. Delivering on these opportunities is particularly important given the expectation among industry actors that travel plans will continue to be required for new residential developments in the future. Should these opportunities not be realised, existing issues will continue to remain, thereby limiting the potential of travel planning in the context of new residential developments.

Second, as part of enhancing the impacts of travel plans for new residential developments, local government will need to become more involved in the travel planning process. A particular focus will need to be placed on undertaking a more systematic assessment of travel plan quality prior to granting planning approval. This is likely to require coordination between different administrative units within councils, such as statutory planning and transport/traffic engineering, to ensure suitable input is provided at the planning application stage. A shift towards a more pro-active culture of planning enforcement will also need to take place. This is particularly relevant in the context of travel plans for new residential developments given that a lack of implementation may not necessarily result in public complaints.

Third, building the capacity of the industry to deliver effective travel planning will take considerable time which needs to be acknowledged. This will require strong leadership from government to develop sound planning requirements backed by sufficient opportunities for training, including the provision of clear guidelines. Training and guidelines will need to give sufficient attention to implementation, given that this step in the travel planning process is typically faced with greater difficulties than other aspects. Greater involvement from property managers in implementation, and preferably also in preparing residential travel plans, will also take considerable time given that transport has not traditionally featured as a core function of their business.

Fourth, in the absence of suitable control sites, caution will need to be applied when using only secondary data as a comparator for evaluating the impacts of travel plans for new residential developments. Such comparisons may overestimate impacts due to inconsistencies in geographical locations and data collection periods, as evidenced by Chap. 8. Therefore, where future evaluations are limited to secondary data comparisons, this will need to be clearly acknowledged. Furthermore, the potential for self-selection to contribute to observed differences in travel behaviour will also need to be acknowledged, if not controlled for, to avoid overstating the impacts of travel plans at new residential developments.

Finally, given the finding that travel plans can contribute to significantly lower car use at new residential developments, their role should be recognised as an important element of TDM and transport policy more generally. This is particularly relevant for cities experiencing transport pressures associated with increasing demand for new housing developments.

10.4 Critique

While this thesis has provided a number of original contributions to knowledge, it is also subject to a number of limitations.

In considering the survey of councils reported in Chap. 5, this was limited to the state of Victoria despite a lack of understanding concerning the scale of travel planning practice for new developments in other states of Australia, as well as other countries. Furthermore, the survey did not include a question about the types of land use that travel plans have been required for. In hindsight, this information would have been valuable given the focus of subsequent chapters that were specific to new residential developments.

While best efforts were made to interview a range of industry representatives, only three property developers and three property managers were able to be recruited, as reported in Chap. 6. These actors proved particularly difficult to recruit as transport is not seen as a core function of their business. Including more property developers and property managers in the sample would have most likely provided richer insight on the perspectives of these actors.

The framework used to assess the quality of the travel plans, as reported in Chap. 7, was developed solely by the researcher. While this was based on best practice elements reported in the literature, some practitioners may be of the opinion that certain criteria should be assigned higher (or lower) implied weightings than those specified. However, given the focus was on the relative strengths of the travel plans and their areas for improvement, rather than the final score alone, this limitation is not considered to present any major issues. Another limitation however was that the travel plans sourced for the assessment may not have been representative of all travel plans in terms of their quality.

The case studies reported in Chap. 8 involved only four new residential developments with travel plans. This was due to the considerable amount of resources required for data collection at each of the case and control sites. In addition, the resident travel survey did not attract a sufficient sample size for statistical power requirements to be met. This limited the ability to make any definitive statements about the magnitude of the self-selection effect.

Overall, exploring each research component in greater detail would have provided additional insight on the research findings. However, doing so within the resources available would have compromised the research scope, potentially leading to a smaller number of research components. For example, councils from other states in Australia could have been included in the survey to offer a broader perspective and allow for cross-jurisdiction comparisons but this may have meant not undertaking any interviews with industry representatives, thereby limiting the understanding of other actors' perspectives. Similarly, assessing the quality of a greater number of travel plans would have provided a greater appreciation of their relative strengths and areas for improvement, but may have resulted in having to reduce the scope of the case studies, thereby limiting the understanding of travel plan impacts.

A final limitation is the geographical context in which this research was undertaken. The research findings are based on experience from the Australian state of Victoria and may therefore differ in other jurisdictions. In particular, the site-specific nature of travel planning means that the local context should always be considered. Despite this, the findings have wider geographical implications given that travel plans are used in other states and countries.

10.5 Future Research Directions

Based on the limitations identified in the previous section, it is now possible to suggest a number of avenues for future research.

Examining the scale of travel planning practice for new developments in other states of Australia and other countries would help to provide a broader perspective of current practice and offer additional insight through cross-jurisdiction comparisons. The research method detailed in Chap. 5 could be replicated for this purpose. However, it would be desirable to include an additional question to determine the number of travel plans required by land use type. Furthermore, given that different planning systems are used in each state of Australia and in other countries, the planning context would also need to be taken into account when assessing and comparing results across jurisdictions.

There is a need to also develop a stronger understanding of the perspectives of property developers and managers involved in travel planning for new residential developments. These stakeholders represent key actors in the process so it is important that their perspectives are well understood so that future practice can be further enhanced.

More could also be done to refine the framework used for assessing travel plan quality by taking into account the collective views of travel planning practitioners, perhaps through the use of a 'Delphi' survey. This would help to ensure that the criteria and scoring system used in the assessment framework aligns with the preferences of industry representatives involved in travel planning. Assessing the quality of travel plans prepared in other jurisdictions would also be useful for gaining a stronger understanding of best practice.

Another key area for future research involves building up a stronger evidence base of the effectiveness of travel plans for new residential developments, not only in the state of Victoria but also in other jurisdictions across Australia and internationally. Where possible, control sites should be used as comparators in preference to secondary data to avoid overstating the impacts of travel plans. Related to this is the need to develop a stronger quantitative understanding of the extent of self-selection associated with travel plans for new residential developments. Surveys at additional sites would help to provide a larger sample size so that estimates of self-selection can be made with a greater level of statistical confidence.

While not based on a specific limitation of this study, future research could also look to assess the relative merits of different approaches for implementing travel plans at new residential developments. This would help to establish an understanding of which methods of implementation are most appropriate under different contexts.

There is also a need to empirically test the integrated theory of implementation and enforcement by applying it in practice to travel planning for new residential developments. The integrated theory could also be used more broadly by applying it to other sectors, such as those where minimum standards are crucial.

In closing, this thesis has explored the use of travel plans for new residential developments by assessing their effectiveness and identifying opportunities to enhance their effectiveness. Acting on these opportunities will help to improve the way in which travel plans are developed, implemented and monitored at new residential developments into the future, ultimately supporting a greater uptake of more sustainable forms of transport.

Appendix A
Common Travel Plan Measures

Travel plan measure	Description/comment
Walking	
Home zone features	Shared spaces with traffic calming measures, particularly relevant for residential travel plans (Department for Transport 2005)
On-site facilities and services	Examples include banking facilities, post offices, cafes and shops. Reduces the need to travel by car by facilitating walking to local destinations (Giuliano et al. 1991)
Pedestrian infrastructure improvements	Examples include upgraded footpaths, new links, crossing facilities and improved lighting (Cairns and Newson 2006; Coleman 2000; Woodruff and Hui 2010)
Pedometer program	Often linked to a website to track daily steps. Most suitable for workplaces
Umbrellas for rainy days	Can facilitate walking over car use for short trips (Cairns et al. 2010)
Walking school bus	Generally limited to primary school students (Peddie and Somerville 2005)
Cycling	
Bicycle couriers	Can reduce amount of car/truck related deliveries (MAX 2009a)
Bicycle end of trip facilities	Includes secure parking, showers, changing rooms and lockers (Rye 2002b)
Bicycle fleet or bicycle share scheme	Applicable to workplaces, universities and residential sites (Balsas 2003; Wiblin 2010)
Bicycle loan scheme	Interest free loans or free bicycles to participants (Cairns et al. 2010; Rye 2002b)
Bicycle repair service	Either on-site or mobile service (Cairns et al. 2010)

(continued)

C. De Gruyter, *Travel Plans for New Residential Developments: Insights from Theory and Practice*, Springer Theses,
DOI 10.1007/978-981-10-2092-6

(continued)

Travel plan measure	Description/comment
Bicycle training	Relevant to schools (Hansen et al. 2012) and workplaces (Myers 2005)
Bicycle users group	Relevant to workplaces and residential sites (Coleman 2000; Harrison 2003)
Cycling infrastructure improvements	Examples include new/upgraded cycling paths, new on-road markings, crossing facilities and improved lighting (Cairns and Newson 2006; Coleman 2000; Woodruff and Hui 2010)
Public transport	
Discounted or free public transport tickets	Common in both workplaces and residential sites (Myers 2005; Rye 1999b), but also found in universities (Cooper and Meiklejohn 2003; Curtis and Holling 2004)
New or upgraded public transport services	Usually only found at relatively large sites and typically funded through developer contributions for new developments. May include stop upgrades at smaller sites
Shuttle bus	Common in workplaces, often as a link to nearby rail services (Cairns et al. 2010)
Car parking	
Parking cash-out	Typically limited to workplaces where an employee can receive the cash value of a parking space in lieu of using that parking space (Green 1995; Potter et al. 1999)
Parking charges and restrictions	Commonly applied to restrict car use. Found to work well in combination with financial incentives to use other modes (Cairns et al. 2010; Coleman 2000; Rye and Ison 2005)
Parking permit policies	Common example involves not providing parking permits for those living close to the site (Cairns et al. 2010; Rye and Ison 2005; Wake et al. 2010)
Unbundled parking	Parking sold separately to a home or office (for example) to better reflect parking demand (Department for Transport 2005)
Use of revenue raised from parking charges	Example includes 'ring-fencing' the funds obtained and using these for implementing sustainable transport measures as part of a travel plan (Rye and Ison 2005)
Carpooling	
Carpool matching program	Common in workplaces and universities (Balsas 2003; DeGruyter et al. 2005)
Discounted or free parking for carpoolers	Usually limited to workplaces and universities (Giuliano et al. 1991)

(continued)

(continued)

Travel plan measure	Description/comment
Guaranteed ride home program	Such programs rarely used in practice but highly valued as a form of 'insurance' if ride home falls through (Berman and Radow 1997; Coleman 2000)
Priority carpool parking	Usually limited to workplaces and universities (Cairns et al. 2010; Thom 2009)
Car sharing	
Car sharing service	Common in residential developments to reduce car ownership but also applicable to workplaces for business related trips (Department for Transport 2005; Wiblin 2010)
Free or discounted car sharing membership	Normally limited to residential sites to encourage the uptake of car sharing
Marketing and promotion	
Events	Examples include a travel plan launch as well as regular events such as Ride to Work Day (Hinckson and Badland 2011; Wiblin 2010; Woodruff and Hui 2010)
Induction sessions	Transport information typically provided for new staff at workplaces (Department for Transport 2005)
Information provision	Examples include public transport timetables, maps and websites (Cairns et al. 2010)
Marketing and promotion	General marketing of the travel plan and its benefits (Cairns et al. 2010; Rye 2002b)
Welcome packs	For new residents or employees, with information on transport options as well as incentives, e.g. free public transport ticket (Cooper and Meiklejohn 2003; Myers 2005)
Financial incentives	
Reward schemes	For users of more sustainable transport modes (Myers 2005; Potter et al. 1999)
Sustainable transport allowances	Examples include bicycle and public transport mileage allowance (Cairns et al. 2010; Coleman 2000)
Tax incentives	Can reduce cost of using more sustainable transport modes (Enoch and Rye 2006)
Discounts at local retailers	Encourages local shopping, particularly by walking (Balsas 2003; Enoch and Rye 2006)
Vanpool subsidies	Funding to assist with ongoing operation of vanpools (Balsas 2003; Enoch and Rye 2006)
Travel plan management	
Local travel plan group membership	Can provide a supportive network for the delivery and monitoring of travel plans (Enoch 2012a; Enoch et al. 2007; Ratio Consultants Pty Ltd 1991)

(continued)

(continued)

Travel plan measure	Description/comment
Travel plan coordinator	Recognised as a key component of any travel plan (Balsas 2003; Wiblin et al. 2012)
Travel plan working group	Provides support to the travel plan coordinator in the delivery and monitoring of the travel plan and assists with establishing organisational commitment (Enoch 2012a)
Working practices	
Teleconferencing facilities	Can eliminate the need for some work related trips (Enoch 2012b)
Videoconferencing facilities	Can eliminate the need for some work related trips (Enoch 2012b; Wake 2012)
Flexitime	Time in lieu which can reduce commuting trips (Giuliano et al. 1991; Potter et al. 1999)
Telecommuting	Working from home which can reduce commuting trips (Cairns et al. 2010; Rye 1999b)
Compressed working weeks	Common example is 80 h worked in 9 days, with 1 leave day taken per fortnight. Can reduce commuting trips (Giuliano et al. 1991; Potter et al. 1999)
Staggered work hours	Can reduce localised congestion, e.g. at key entrance/exit points (Giuliano et al. 1991)
Other	
Curriculum program	Education about travel plans as well as road safety (Cairns and Newson 2006; Di Pietro and Hughes 2003; Howlett and Watson 2010; Peddie and Somerville 2005)
Early bell for students	Students who walk, cycle or catch public transport can leave school 10 min early (Peddie and Somerville 2005)
Eco-driving courses	Promotes smoother driving practices and less fuel use (MAX 2009b)
Lobbying for infrastructure or service improvements	Travel plan provides strong evidence base for improvements needed (Cairns et al. 2010; Rye 2002b; Wiblin et al. 2012)

Source Author's synthesis of the literature based on citations within the table
Note Measures were placed into one category only despite some being applicable to more than one category

Appendix B
Success Factors for Travel Plans

Success factor	Relative importance	Comments
Ownership and engagement	Very high	Often achieved through the target group developing the travel plan in accordance with their own directions (Cairns and Newson 2006; Howlett and Watson 2010)
Senior management support	Very high	Includes leading by example, particularly in workplaces (Baudains 2003; Cairns et al. 2010; Rye 1997) and schools (Newson et al. 2010)
Enthusiastic and dedicated travel plan coordinator	Very high	Widely cited in the literature (Hendricks 2005; Hendricks and Georggi 2007; Rye 1997; Van Malderen et al. 2013)
Comprehensive travel plan measures	Very high	Measures that work together as an integrated package and are tailored to the needs of the site (Cairns et al. 2010; Ison and Rye 2008; MAX 2009a)
Constraints on car parking	Very high	Widely supported by the literature (Balsas 2003; Bianco 2000; Cairns et al. 2010; Hamre and Buehler 2014)
Supportive policy framework	Very high	Particularly relevant when requiring travel plans for new developments (Addison & Associates 2008; Department for Transport 2005; Enoch 2012a).

(continued)

C. De Gruyter, *Travel Plans for New Residential Developments: Insights from Theory and Practice*, Springer Theses,
DOI 10.1007/978-981-10-2092-6

(continued)

Success factor	Relative importance	Comments
Clear objectives and targets	High	Needs to include agreement on objectives and targets by all relevant parties (Addison 2002; Cairns and Newson 2006; Cairns et al. 2010; Wake et al. 2010)
Partnerships	High	Particularly with relevant government agencies (ATOC 2013; Black and Schreffler 2010; Woodruff and Hui 2010)
Demonstration of benefits	High	Particularly relevant to employers (DeGruyter et al. 2005; Roby 2010)
Combination of incentives and disincentives	High	Also referred to as 'carrots' and 'sticks' or 'push' and 'pull' measures (Addison 2002; Cairns et al. 2010, 2004; Enoch and Rye 2006)
Dedicated funding	High	Particularly for implementation purposes, but also for monitoring (Baslington 2008; Cairns et al. 2004; Davison et al. 2010; Wiblin et al. 2012).
Clear roles and responsibilities	High	Particularly important for implementation purposes (Addison 2002; Department for Transport 2008; Wake et al. 2010)
Tax incentives	High	Key example is the removal of tax penalties on employer subsidies for travel plan measures (Davison et al. 2010; Potter et al. 1999; Rye 1999a).
Appropriate monitoring techniques	High	Examples include: using consistent survey methods and questions to enable valid comparisons (Ampt et al. 2009); taking direct measurements such as bicycle and car parking counts (Sullivan and Percy 2008); monitoring 'process' factors such as participation rates and awareness levels (Wake 2012); and in the case of new developments, linking monitoring to financial sanctions (Grant et al. 2012)

(continued)

(continued)

Success factor	Relative importance	Comments
Working groups	Moderate	Can assist the travel plan coordinator with implementation and help to establish commitment from various parts of an organisation (Baudains 2003; Rye 1997)
Local travel plan groups	Moderate	Provides forum for capacity building, networking, training and sharing best practice (Enoch et al. 2007; Tyler et al. 2007)
Flexibility	Moderate	The travel plan should be a 'living' document and ongoing process, particularly for new developments where the occupant/s may be unknown (Addison 2002)
'Hard' infrastructure measures	Moderate	Helps to complement and 'lock-in' the benefits of travel plans (Cairns et al. 2008)
Training, guidelines and resources	Moderate	Important that these are location-specific and tailored to the type of travel plan being considered (Addison 2002; Department for Transport 2002)
Travel plan introduced early	Moderate	Relevant to new developments; concept should be introduced before the planning application is lodged (Department for Transport 2005, 2007; MAX 2009a)
Integration with transport assessments	Moderate	Relevant to new developments (East Sussex County Council 2008; Fraser and Addison 2002; Transport for London 2010)
Assessment of transport context	Moderate	Needs to consider aspects both external (e.g. public transport services) and internal (e.g. workplace travel policies) to the site (Smith 2010; Wake et al. 2010)
Economic development	Moderate	Applies mostly to new developments (Rye 2002a)
Media coverage/other recognition	Low	Can assist in enhancing motivation, triggering new activity and sustaining interest (Cairns and Newson 2006)

Source Author's synthesis of the literature based on citations within the table
Note 'Relative importance' rating based on the author's assessment of the literature; it is acknowledged that this rating may vary depending on local circumstances

Appendix C
Council Survey Questionnaire

1. To what extent do you agree that the following mechanisms are effective in managing transport access for new developments?

Mechanism	Strongly disagree	Disagree	Neutral	Agree	Strongly agree
Upgrading the surrounding road network and/or intersections	☐	☐	☐	☐	☐
Providing sufficient car parking to meet demand	☐	☐	☐	☐	☐
Providing new and/or improved public transport services	☐	☐	☐	☐	☐
Providing a safe and connected walking and cycling network	☐	☐	☐	☐	☐
Incorporating a mix of land uses to potentially reduce the length and number of trips	☐	☐	☐	☐	☐
Developing and implementing a travel plan to encourage the use of more sustainable transport modes	☐	☐	☐	☐	☐
Other (please state): ____________________	☐	☐	☐	☐	☐

2. Has your council ever required a travel plan for any type of new development?

Yes	☐
No (**go to question 7**)	☐
Unsure (**go to question 8**)	☐

C. De Gruyter, *Travel Plans for New Residential Developments: Insights from Theory and Practice*, Springer Theses,
DOI 10.1007/978-981-10-2092-6

3. How many new developments has your council required a travel plan for in the last two years?

None	☐
1–2	☐
3–5	☐
6–10	☐
More than 10	☐
Unsure	☐

4. For any new developments in your municipality that have required a travel plan, what level of monitoring has taken place to determine whether actions in the travel plan are being implemented?

Not applicable as no implementation has taken place yet	☐
No monitoring has taken place yet and there are no plans to do any monitoring	☐
No monitoring has taken place yet but there may be some monitoring in the future	☐
No monitoring has taken place yet but there will definitely be some monitoring in the future	☐
Some travel plans have been monitored but the majority have not been monitored yet	☐
Most travel plans have been monitored	☐
All travel plans have been monitored to date	☐
Unsure	☐

5. Which of the following mechanisms have been used by your council to require a travel plan for a new development (tick all that apply)?

Condition on planning permit	☐
Section 173 agreement	☐
Development contribution plan	☐
Verbal negotiation with developer	☐
Other (please state): ____________________	☐
Unsure	☐

6. What are the key reasons/motivations for your council requiring travel plans for new developments (tick all that apply)?

Mitigate transport impacts and improve accessibility (**go to question 8**)	☐
Reduce requirements for road network upgrades (**go to question 8**)	☐
Offset impacts of providing reduced levels of car parking (**go to question 8**)	☐
Contribute to wider environmental objectives (**go to question 8**)	☐
Provide a selling feature for new developments (**go to question 8**)	☐
Deliver on council's transport policies and/or strategies (**go to question 8**)	☐
Other (**go to question 8**)	☐
Unsure (**go to question 8**)	☐

7. Why has your council never required a travel plan for any new development (tick all that apply)?

Little or no awareness of the travel plan concept	☐
Travel plans not considered to be effective or appropriate for the local area	☐
Planning to require travel plans for new developments in the future	☐
Other (please state): ______________________________	☐
Unsure	☐

8. Which of the following best describes your level of familiarity and experience with travel plans?

I had never heard about travel plans prior to this survey (**go to question 10**)	☐
I have heard about travel plans before but do not understand how they work in practice	☐
I am aware of travel plans but only have a limited understanding of how they work in practice	☐
I am very aware of what travel plans are but do not have any practical experience in using them	☐
I am very aware of what travel plans are and have practical experience in using them	☐

9. What is the future likelihood that your council will require a travel plan for a new development in the next 12 months?

Highly unlikely	☐
Unlikely	☐
Unsure	☐
Likely	☐
Highly likely	☐

10. For cross-classification purposes only, please indicate the location of your council:

Inner metropolitan (Maribyrnong, Melbourne, Port Phillip, Stonnington, Yarra)	☐
Middle metropolitan (Banyule, Bayside, Boroondara, Darebin, Glen Eira, Hobsons Bay, Kingston, Manningham, Monash, Moonee Valley, Moreland, Whitehorse)	☐
Outer metropolitan (Brimbank, Cardinia, Casey, Greater Dandenong, Frankston, Hume, Knox, Maroondah, Melton, M'ton Peninsula, Nillumbik, Whittlesea, Wyndham, Yarra Ranges)	☐
Regional (any council not listed in the above categories)	☐

11. If you have any other comments in relation to transport planning for new developments, particularly with respect to travel plans, please provide them in the space below:

Appendix D
Letter to Property Managers

MONASH University

Institute of Transport Studies (Monash)
The Australian Research Council Key Centre in Transport Management
Department of Civil Engineering
Faculty of Engineering

28 February 2014

[redacted]

Dear [redacted]

Research into travel characteristics of residents living in apartment buildings in Melbourne

The Institute of Transport Studies at Monash University is currently undertaking research into the travel characteristics of residents living in apartment buildings in Melbourne, as part of a PhD research project. Only eight apartment buildings in Melbourne, including [redacted], have been selected for this research due to their unique characteristics.

We would therefore like to request your assistance with this research which will provide a significant contribution towards the future planning of residential developments. This assistance includes:

- **Allowing us to conduct a voluntary, anonymous travel survey of residents via a letter box drop:** we are proposing to hand deliver information about the survey to each letter box during April or May 2014. The information to be delivered will include a letter, followed by a survey postcard (three days later) and a reminder postcard (one week after the survey postcard). As a token of our appreciation, residents that choose to participate in the survey will have the chance to win a $250 Coles Myer gift card. Contact details will only be sought for those residents who elect to go into the prize draw. These contact details will be deleted prior to any data analysis to ensure that **all survey responses are anonymous.** The survey has been approved by the Monash University Human Research Ethics Committee (Project Number CF12/1205 – 2012000586).
- **Allowing us to measure car and bicycle parking utilisation in the building:** this will involve us counting the number of cars and bicycles parked in the building between 6am and 7am on a Tuesday, Thursday and Saturday within the same single week during April or May 2014. The count will take around 10 minutes to complete each morning. A building manager or caretaker could accompany us to undertake the count or access could be arranged through the provision of a security pass or key.

In addition to the above, we will also be conducting a traffic count in the local area over a one week period in April or May 2014 to act as a point of reference for our research.

All data that is collected and analysed by us will only be reported in aggregate format and will not, under any circumstances, contain individual names or other potentially identifiable characteristics.

Following the completion of our research, we will provide you with a summary of our findings. Your assistance with supporting this research is greatly appreciated. If you have any questions, please do not hesitate to contact Chris De Gruyter, PhD Candidate on 0403 073 743 or chris.degruyter@monash.edu.

Yours sincerely,

GRose

Professor Geoff Rose
Institute of Transport Studies, Monash University

Postal - Monash University, VIC 3800, Australia
Building 60, Clayton Campus, Wellington Road, Clayton
Telephone +61 3 9905 9627 Facsimile +61 3 9905 9493
Email itsinfo@eng.monash.edu.au eng.monash.edu.au/civil/research/centres/its/
CRICOS Provider No. 00008C ABN 12 377 614 012

ITS (Monash)

C. De Gruyter, *Travel Plans for New Residential Developments: Insights from Theory and Practice*, Springer Theses,
DOI 10.1007/978-981-10-2092-6

Appendix E
Pre-Notification Letter

MONASH University

Institute of Transport Studies (Monash)
The Australian Research Council Key Centre in Transport Management
Department of Civil Engineering
Faculty of Engineering

May 2014

Understanding how you travel will help shape future transport planning for residential apartment buildings

A PhD research project being undertaken through the Institute of Transport Studies at Monash University aims to improve future transport planning for residential apartment buildings.

As part of the project, we are undertaking a short, anonymous survey to better understand the travel characteristics of people living in residential apartment buildings. Households in selected apartments across Melbourne are being invited to participate. Your response will have a significant impact on future transport planning for residential apartment buildings.

In a few days, you will receive details in the mail about the travel survey.

The survey will only take 10 minutes to complete and as a sign of our appreciation for your participation, **we are offering you the chance to win a Coles Myer gift card worth $250.**

Should you have any questions about the survey process, please contact Chris De Gruyter, PhD Candidate, on 0403 073 743.

Yours sincerely,

G. Rose

Professor Geoff Rose
Institute of Transport Studies, Monash University

Postal - Monash University, VIC 3800, Australia
Building 60, Clayton Campus, Wellington Road, Clayton
Telephone +61 3 9905 9627 Facsimile +61 3 9905 9493
Email itsinfo@eng.monash.edu.au
eng.monash.edu.au/civil/research/centres/its/

Institute of Transport Studies (Monash)

CRICOS Provider No. 00008C ABN 12 377 614 012

C. De Gruyter, *Travel Plans for New Residential Developments: Insights from Theory and Practice*, Springer Theses,
DOI 10.1007/978-981-10-2092-6

Appendix F
Resident Survey Postcard

Front

Understanding how you travel will help shape future transport planning for residential apartment buildings

Help shape future transport planning for residential apartments

Recently we sent your household a letter inviting you to participate in an anonymous travel survey. The survey is investigating the travel characteristics of people living in residential apartment buildings in Melbourne. Your response will have a significant impact on future transport planning for residential apartment buildings.

It only takes 10 minutes and you will have the chance to win a Coles Myer gift card worth $250.

https://www.surveymonkey.com/s/CA1B

MONASH University

AUSTRALIA ▸ CHINA ▸ INDIA ▸ ITALY ▸ MALAYSIA ▸ SOUTH AFRICA

GROUP OF EIGHT

CRICOS provider: Monash University 00008C

Back

Please use the link below or the QR code to go to the survey. Household members 5 years of age and older are invited to complete the survey. It takes about 10 minutes to complete the survey.

Survey participants who respond by 18 May 2014 will have the chance to win a Coles Myer gift card worth $250.

If you prefer a paper version of the survey, or if you have any questions, please contact Chris De Gruyter, PhD Candidate at the Institute of Transport Studies, on 0403 073 743.

Simply scan the QR code or visit: **https://www.surveymonkey.com/s/CA1B**

MONASH University

AUSTRALIA ▸ CHINA ▸ INDIA ▸ ITALY ▸ MALAYSIA ▸ SOUTH AFRICA

Institute of Transport Studies (Monash)

To protect the integrity of data that is collected, only one survey can be completed per computer or smart phone device. Any other household members completing the survey will need to do so using a different computer or smart phone device. Alternatively, a paper version of the survey can be requested from Chris De Gruyter, PhD Candidate at the Institute of Transport Studies, on 0403 073 743.

C. De Gruyter, *Travel Plans for New Residential Developments: Insights from Theory and Practice*, Springer Theses,
DOI 10.1007/978-981-10-2092-6

Appendix G
Resident Survey Questionnaire

1. How many people usually live in your household, including yourself?

Number of people	☐

2. How long have you been living at your current address?

Years	☐
Months	☐

3. Do you currently rent at your present address?

Yes	☐
No	☐

4. What is your current employment status? Select one option only.

Employed full-time (35 h or more per week)	☐
Employed part-time (less than 35 h per week)	☐
Employed on a casual basis	☐
Not currently employed (**go to question 6**)	☐

C. De Gruyter, *Travel Plans for New Residential Developments: Insights from Theory and Practice*, Springer Theses,
DOI 10.1007/978-981-10-2092-6

5. Since you have been living at your current address, how do you travel to your current workplace/s? Please provide a response for each method of transport.

Method of transport	Never	Rarely	Occasionally	Sometimes	Mostly	Almost always
	0 % of the time	Up to 20 % of the time	20–40 % of the time	40–60 % of the time	60–80 % of the time	80–100 % of the time
Public transport (train, tram or bus)	☐	☐	☐	☐	☐	☐
Walk only	☐	☐	☐	☐	☐	☐
Car—as the driver	☐	☐	☐	☐	☐	☐
Car—as a passenger	☐	☐	☐	☐	☐	☐
Bicycle	☐	☐	☐	☐	☐	☐
Motorcycle or scooter	☐	☐	☐	☐	☐	☐
Taxi	☐	☐	☐	☐	☐	☐
Aeroplane	☐	☐	☐	☐	☐	☐
Do not travel (e.g. work from home)	☐	☐	☐	☐	☐	☐
Other	☐	☐	☐	☐	☐	☐

6. Are you currently enrolled at an educational institution (e.g. school, university, TAFE)?

No (**go to question 8**)	☐
Yes, primary or secondary school	☐
Yes, full-time at university or TAFE	☐
Yes, part-time at university or TAFE	☐
Yes, at a different educational institution	☐

7. Since you have been living at your current address, how do you travel to your current educational institution/s? Please provide a response for each method of transport.

Method of transport	Never	Rarely	Occasionally	Sometimes	Mostly	Almost always
	0 % of the time	Up to 20 % of the time	20–40 % of the time	40–60 % of the time	60–80 % of the time	80–100 % of the time
Public transport (train, tram or bus)	☐	☐	☐	☐	☐	☐
Walk only	☐	☐	☐	☐	☐	☐
Car—as the driver	☐	☐	☐	☐	☐	☐
Car—as a passenger	☐	☐	☐	☐	☐	☐
Bicycle	☐	☐	☐	☐	☐	☐
Motorcycle or scooter	☐	☐	☐	☐	☐	☐
Taxi	☐	☐	☐	☐	☐	☐
Aeroplane	☐	☐	☐	☐	☐	☐
Do not travel (e.g. online course)	☐	☐	☐	☐	☐	☐
Other	☐	☐	☐	☐	☐	☐

8. Since you have been living at your current address, how do you travel when you go to the shops (e.g. supermarket, bakery, other retail)? Please provide a response for each method of transport.

Method of transport	Never	Rarely	Occasionally	Sometimes	Mostly	Almost always
	0 % of the time	Up to 20 % of the time	20–40 % of the time	40–60 % of the time	60–80 % of the time	80–100 % of the time
Public transport (train, tram or bus)	☐	☐	☐	☐	☐	☐

Method of transport	Never	Rarely	Occasionally	Sometimes	Mostly	Almost always
	0 % of the time	Up to 20 % of the time	20–40 % of the time	40–60 % of the time	60–80 % of the time	80–100 % of the time
Walk only	☐	☐	☐	☐	☐	☐
Car—as the driver	☐	☐	☐	☐	☐	☐
Car—as a passenger	☐	☐	☐	☐	☐	☐
Bicycle	☐	☐	☐	☐	☐	☐
Motorcycle or scooter	☐	☐	☐	☐	☐	☐
Taxi	☐	☐	☐	☐	☐	☐
Do not travel (e.g. online shopping)	☐	☐	☐	☐	☐	☐
Other	☐	☐	☐	☐	☐	☐

9. How many cars are owned or used by your household? *Exclude any car share vehicles (e.g. Go Get)*

Number of cars (**if none, go to question 11**)	☐

10. Where are these car/s usually parked at night? (Response options tailored to site context)

Within the residential building car park	☐
On the street	☐
Commercial car park	☐
Other location	☐

11. How many bicycles are owned or used by your household?

Exclude any bicycles that are part of Melbourne Bike Share

Number of bicycles (**if none, go to question 13**)	☐

12. Where are these bicycle/s usually parked at night? (Response options tailored to site context)

Common bicycle storage area	☐
Individual storage cage	☐
Car park space	☐
Within your residential unit or balcony	☐
Outside the building somewhere	☐
Other location	☐

13. To what extent do you agree with the following statements?

Statement	Strongly disagree	Disagree	Neutral	Agree	Strongly agree
I prefer to travel by car—whenever possible	☐	☐	☐	☐	☐
It is important that I have my own allocated car parking space at home	☐	☐	☐	☐	☐
Others might think I had a financial difficulty if I did not have a car	☐	☐	☐	☐	☐
I prefer to take public transport rather than travel by car—whenever possible	☐	☐	☐	☐	☐
I prefer to use a bicycle rather than travel by car—whenever possible	☐	☐	☐	☐	☐
I like to live where I have shops within walking distance of my home	☐	☐	☐	☐	☐

14. Please indicate your awareness and use of the following travel initiatives which may be available at your residential building. (Response options tailored to site context)

Travel initiative	Not aware of it or doesn't exist	Aware, but haven't used it	Aware and have used it
Common bicycle storage area	☐	☐	☐
Bicycle fleet in building car park	☐	☐	☐

Travel initiative	Not aware of it or doesn't exist	Aware, but haven't used it	Aware and have used it
Free membership to Melbourne bike share	☐	☐	☐
Go get car sharing vehicle in building car park	☐	☐	☐
Free membership to go get car share	☐	☐	☐
Transport information on building's website	☐	☐	☐
Online resident forum for organising carpooling	☐	☐	☐
Transport information in new residents kit	☐	☐	☐
Free public transport tickets for new residents	☐	☐	☐
Umbrellas at reception/lobby area	☐	☐	☐

15. How useful would you find the following travel initiatives at your residential building? (Response options tailored to site context)

Statement	Not at all useful	Slightly useful	Moderately useful	Very useful	Extremely useful
More car parking spaces	☐	☐	☐	☐	☐
Car parking sold/rented separately to reduce costs for those not needing a parking space	☐	☐	☐	☐	☐
A shared car that can be rented out by the hour (maintained by a third party)	☐	☐	☐	☐	☐

Statement	Not at all useful	Slightly useful	Moderately useful	Very useful	Extremely useful
A discounted public transport ticket (Myki)	☐	☐	☐	☐	☐
More locations to safely store your bicycle/s	☐	☐	☐	☐	☐
An on-site bicycle maintenance service	☐	☐	☐	☐	☐
An information screen in the foyer displaying current tram, train and bus departure times	☐	☐	☐	☐	☐
Information about your transport options, including timetables and maps	☐	☐	☐	☐	☐

To enable us to compare our sample with population statistics obtained from the most recent Census (2011), we would like you to answer the same income question that was asked in the Census.

16. What was the total of all wages/salaries, government benefits, pensions, allowances and other income that you usually receive? Select one option only. *Do not deduct: tax, superannuation contributions, health insurance, amounts salary sacrificed, or any other automatic deductions.*

\$2,000 or more per week (\$104,000 or more per year)	☐
\$1,500–\$1,999 per week (\$78,000–\$103,999 per year)	☐
\$1,250–\$1,499 per week (\$65,000–\$77,999 per year)	☐
\$1,000–\$1,249 per week (\$52,000–\$64,999 per year)	☐
\$800–\$999 per week (\$41,600–\$51,999 per year)	☐
\$600–\$799 per week (\$31,200–\$41,599 per year)	☐

$400–$599 per week ($20,800–$31,199 per year)	☐
$300–$399 per week ($15,600–$20,799 per year)	☐
$200–$299 per week ($10,400–$15,599 per year)	☐
$100–$199 per week ($1–$10,399 per year)	☐
Nil income	☐
Negative income	☐

17. What is your age?

19 years or less	☐
20–29 years	☐
30–39 years	☐
40–49 years	☐
50–59 years	☐
60–69 years	☐
70 years or more	☐

18. Are you male or female?

Male	☐
Female	☐

19. Do you have any comments in relation to travel in your local area?

If you would like to go into the draw to win a Coles Myer gift card worth $250, please provide your first name and a contactable phone number OR email address below. Please note that these details will not, under any circumstances, be used for any purpose other than contacting the successful prize winner.

First name	
Phone number OR email address	

Thank you for taking the time to complete this survey. The results will help shape future transport planning for residential apartment buildings.

Please place your completed survey in the reply paid envelope provided and ensure this is post marked by 18 May 2014.

References

Addison, L. (2002). *Using the planning process to secure sustainable transport.* Paper presented to European Transport Conference, Homerton College, Cambridge, UK.

Addison & Associates. (2008). *Delivering travel plans through the planning process—Research report.* London, UK: Department for Transport and Communities and Local Government.

Ampt, E. S., Richardson, A. J., & Wake, D. (2009). *Simple and suited: Guidelines for workplace travel surveys.* Paper presented to 32nd Australasian Transport Research Forum (ATRF), Auckland, New Zealand.

ATOC. (2013). *Guidance on the implementation of station travel plans.* UK: Association of Train Operating Companies.

Balsas, C. (2003). Sustainable transportation planning on college campuses. *Transport Policy, 10*, 35–49.

Baslington, H. (2008). School travel plans: Overcoming barriers to implementation. *Transport Reviews, 28*(2), 239–258.

Baudains, C. (2003). *Environmental education in the workplace: Inducing voluntary travel behaviour change to decrease single occupant vehicle trips by commuters into the Perth CBD.* PhD thesis, Murdoch University.

Berman, W., & Radow, L. (1997). Travel demand management in the USA: Context, lessons learned and future directions. *Energy Policy, 25*(14–15), 1213–1215.

Bianco, M. (2000). Effective transportation demand management: Combining parking pricing, transit incentives, and transportation management in a commercial district of Portland, oregon. *Transportation Research Record: Journal of the Transportation Research Board, 1711*, 46–54.

Black, C., & Schreffler, E. (2010). Understanding transport demand management and its role in delivery of sustainable urban transport. *Transportation Research Record: Journal of the Transportation Research Board, 2163*, 81–88.

Cairns, S., & Newson, C. (2006). *Making school travel plans work: Effects, benefits and success factors at English schools.* Association for European Transport and Contributors.

Cairns, S., Newson, C., & Davis, A. (2010). Understanding successful workplace travel initiatives in the UK. *Transportation Research Part A: Policy and Practice, 44*(7), 473–494.

Cairns, S., Sloman, L., Newson, C., Anable, J., Kirkbride, A., & Goodwin, P. (2004). *Smarter choices—Changing the way we travel.* UK: Department for Transport.

Cairns, S., Sloman, L., Newson, C., Anable, J., Kirkbride, A., & Goodwin, P. (2008). Smarter choices: Assessing the potential to achieve traffic reduction using 'soft measures'. *Transport Reviews, 28*(5), 593–618.

Coleman, C. (2000). Green commuter plans and the small employer: An investigation into the attitudes and policy of the small employer towards staff travel and green commuter plans. *Transport Policy, 7*, 139–148.

C. De Gruyter, *Travel Plans for New Residential Developments: Insights from Theory and Practice*, Springer Theses,
DOI 10.1007/978-981-10-2092-6

Cooper, B., & Meiklejohn, D. (2003). *A new approach for travel behaviour change in universities.* Paper presented to 26th Australasian Transport Research Forum (ATRF), Wellington, New Zealand.

Curtis, C., & Holling, C. (2004). Just how (travel) smart are Australian universities when it comes to implementing sustainable travel? *World Transport Policy and Practice, 10*(1), 22–33.

Davison, L., Enoch, M., & Ison, S. (2010). *European experience of travel plans: An expert perspective.* Paper presented to 12th World Conference in Transportation Research, Lisbon, Portugal.

DeGruyter, C., Rao, D., & Meiklejohn, D. (2005). *Tools for travel behaviour change*. Paper presented to 28th Australasian Transport Research Forum (ATRF), Sydney, Australia.

Department for Transport. (2002). *Using the planning process to secure travel plans: Best practice guidance for local authorities, developers and occupiers*. London, UK: Department for Transport.

Department for Transport. (2005). *Making residential travel plans work: Guidelines for new development*, London, UK.

Department for Transport. (2007). *Making residential travel plans work*, London, UK.

Department for Transport. (2008). *The essential guide to travel planning*. London, UK: Department for Transport.

Di Pietro, G., & Hughes, I. (2003). *TravelSMART Schools—There really is a better way to go*! Paper presented to 26th Australasian Transport Research Forum (ATRF), Wellington, New Zealand.

East Sussex County Council. (2008). *Guidance on travel plans for new development.*

Enoch, M. (2012a). *Sustainable transport, mobility management and travel plans.* Surrey, England: Ashgate Publishing Limited.

Enoch, M. (2012b). *Travel plans—opportunities for ICT.* Paper presented to 6th ACM workshop on next generation mobile computing for dynamic personalised travel planning (sense transport'12). Lake District, UK: Low Wood Bay.

Enoch, M., & Rye, T. (2006). Travel plans: Using good practice to inform future policy. In B. Jourquin, P. Rietveld, & K. Westin (Eds.), *Towards better performing transport networks* (pp. 157–177). London, UK: Routledge.

Enoch, M., Zhang, L., & Ison, S. (2007). Unlocking the potential of site based mobility management through local travel plan groups. *World Transport Policy and Practice, 13*(2), 23–39.

Fraser, J., & Addison, L. (2002). Travel plans—A greenwash or an effective planning mechanism. *Town and Country Planning*, 275–277.

Giuliano, G., Hwang, K., Perrine, D., & Wachs, M. (1991). *Preliminary evaluation of regulation XV of the south coast air quality management district*. Berkeley, California, US: University of California Transportation Center.

Grant, M., Rooney, K., & Suter, S. (2012). *Integrated travel demand management into transportation planning: Best practices nationwide.* Paper presented to Transportation Research Board (TRB) 2012 Annual Meeting, Washington DC.

Green, K. (1995). *Looking beyond ECO: Alternatives to employer-based trip reduction.* Policy Study No. 185, Reason Foundation.

Hamre, A., & Buehler, R. (2014). Commuter mode choice and free car parking, public transportation benefits, showers/lockers, and bike parking at work: Evidence from the Washington, DC Region. *Journal of Public Transportation, 17*(2), 67–91.

Hansen, D., Peart, J., & John, G. (2012). *Engaging young minds. Addressing school travel challenges through innovation*. Paper presented to 35th Australasian Transport Research Forum (ATRF), Perth, Australia.

Harrison, J. (2003). Travel plans and the planning system. *Journal of Planning and Environment Law*, 397–403.

Hendricks, S. (2005). Effectiveness of programs for work site trip reduction: The influence of organisational culture. *Transportation Research Record: Journal of the Transportation Research Board, 1924*, 207–214.

Hendricks, S., & Georggi, N. (2007). Documented impact of transportation demand management programs through the case study method. *Journal of Public Transportation, 10*(4), 79–98.
Hinckson, E., & Badland, H. (2011). School travel plans: Preliminary evidence for changing school-related travel patterns in elementary school children. *American Journal of Health Promotion, 25*(6), 368–371.
Howlett, R., & Watson, T. (2010). *Travel planning in Victoria—A new strategic approach to sustaining communities*. Paper presented to 33rd Australasian Transport Research Forum (ATRF), Canberra, Australia.
Ison, S., & Rye, T. (2008). *The implementation and effectiveness of transport demand management measures: An international perspective*. Hampshire, UK: Ashgate.
MAX. (2009a). *WP D—MaxLupo: Guidelines for the integration of mobility management with land use planning*. Sixth Framework Programme, European Commission.
MAX. (2009b). *WP D Integrating mobility management and land use planning: Compendium of site based mobility management measures (D5)*. Sixth Framework Programme, European Commission.
Myers, K. (2005). *Travel behaviour change initiatives: A local government's innovations*. Paper presented to 28th Australasian Transport Research Forum (ATRF), Sydney, Australia.
Newson, C., Cairns, S., & Davis, A. (2010). *Making school travel plans work: Experience from English case studies*, Transport for Quality of Life.
Peddie, B., & Somerville, C. (2005). *Travel behaviour change through school travel planning: Mode shift and community engagement—Results from 33 schools in Victoria*. Paper presented to 28th Australasian Transport Research Forum (ATRF), Sydney, Australia.
Potter, S., Rye, T., & Smith, M. (1999). Tax and green transport plans: A survey of UK experience. *Transport Policy, 6*, 197–205.
Ratio Consultants Pty Ltd. (1991). *Trip reduction ordinance feasibility study*. Australia: VicRoads in association with Greenhouse Unit and Energy Victoria.
Roby, H. (2010). Can travel plans escape the planning Ghetto? *Town and Country Planning*.
Rye, T. (1997). *The implementation of workplace transport demand management in large organisations*. PhD thesis, Nottingham Trent University.
Rye, T. (1999a). Employer attitudes to employer transport plans: A comparison of UK and Dutch experience. *Transport Policy*, *6*, 183–196.
Rye, T. (1999b). Employer transport plans—A case for regulation? *Transport Reviews*, *19*(1), 13–31.
Rye, T. (2002a). *Public-private partnership in securing mobility management through the land-use planning process: A review of examples from the UK and Benelux*. Paper presented to European Conference on Mobility Management (ECOMM), Gent, Belgium.
Rye, T. (2002b). Travel plans: Do they work? *Transport Policy*, *9*, 287–298.
Rye, T., & Ison, S. (2005). Overcoming barriers to the implementation of car parking charges at UK workplaces. *Transport Policy, 12*, 57–64.
Smith, L. (2010). School travel plans: How successful are they? *Traffic Engineering and Control*, 189–193.
Sullivan, C., & Percy, A. (2008). *Evaluating changes associated with workplace and school travel plans—Something old, something borrowed, something new*. Paper presented to 31st Australasian Transport Research Forum (ATRF), Gold Coast, Queensland, Australia.
Thom, A. (2009). *Behaviour change towards sustainable travel in Perth—The TravelSmart workplace program*. Paper presented to PATREC Research Forum, Curtin University Perth, Australia.
Transport for London. (2010). *Transport assessment best practice guidance document*. London, UK: Transport for London.
Tyler, S., Enoch, M., & Zhang, L. (2007). *Setting up local travel plan groups: The future of workplace travel planning in urban conurbations? Examples from London*. Paper presented to 86th Transportation Research Board (TRB) Annual Meeting, Washington DC, US.
Van Malderen, L., Jourquin, B., Pecheux, C., Thomas, I., Van De Vijver, E., Vanoutrive, T., et al. (2013). Exploring the profession of mobility manager in Belgium and their impact on commuting. *Transportation Research Part A, 55*, 46–55.

Wake, D. (2012). *Engaging Perth workplaces for sustainable transport: An evaluation of the TravelSmart workplace program.* Paper presented to 35th Australasian Transport Research Forum (ATRF), Perth, Australia.

Wake, D., Thom, A., & Cummings, R. (2010). *Evaluating workplace travel plans.* Paper presented to 33rd Australasian Transport Research Forum (ATRF), Canberra, Australia.

Wiblin, S. (2010). *Integrating travel behaviour change for workers, shoppers and residents at an outer suburban centre.* Paper presented to 33rd Australasian Transport Research Forum (ATRF), Canberra, Australia.

Wiblin, S., Mulley, C .,& Ison, S. (2012). *Precinct wide travel plans—learnings from Rouse Hill Town Centre.* Paper presented to 35th Australasian Transport Research Forum (ATRF), Perth, Australia.

Woodruff, A., & Hui, C. (2010). *Integrating planning in activity centres: Influencing change across all travel purposes.* Paper presented to 33rd Australasian Transport Research Forum (ATRF), Canberra, Australia.

MIX
Papier aus verantwortungsvollen Quellen
Paper from responsible sources
FSC® C105338

If you have any concerns about our products,
you can contact us on
ProductSafety@springernature.com

In case Publisher is established outside the EU,
the EU authorized representative is:
Springer Nature Customer Service Center GmbH
Europaplatz 3, 69115 Heidelberg, Germany

Printed by Libri Plureos GmbH
in Hamburg, Germany